Borders as Infrastructure

M000207041

Infrastructures Series

edited by Geoffrey C. Bowker and Paul N. Edwards

Borders as Infrastructure

The Technopolitics of Border Control

Huub Dijstelbloem

The MIT Press
Cambridge, Massachusetts
London, England

© 2021 Massachusetts Institute of Technology

This work is subject to a Creative Commons CC-BY-NC-ND license.

Subject to such license, all rights are reserved.

The open access edition of this book was made possible by generous funding from Arcadia—a charitable fund of Lisbet Rausing and Peter Baldwin.

ARCADIA
A charitable fund of Lisbet Rausing and Peter Baldwin

This book was set in Stone Serif and Stone Sans by Westchester Publishing Services. Printed and bound in the United States of America.

Library of Congress Cataloging-in-Publication Data

Names: Dijstelbloem, Huub, 1969– author.
Title: Borders as infrastructure : the technopolitics of border control / Huub Dijstelbloem.
Description: Cambridge, Massachusetts : The MIT Press, [2021] | Series: Infrastructures series | Includes bibliographical references and index.
Identifiers: LCCN 2020045143 | ISBN 9780262542883 (paperback)
Subjects: LCSH: Border security—Technological innovations—Europe. | Infrastructure (Economics)—Europe. | Refugees—Europe. | Europe—Boundaries. | Europe—Emigration and immigration—Government policy. | Europe—Politics and government—1945–
Classification: LCC JV7590 .D55 2021 | DDC 320.1/2—dc23
LC record available at https://lccn.loc.gov/2020045143

10 9 8 7 6 5 4 3 2 1

Contents

Calais shelter, February 2009.
Source: Henk Wildschut.

Preface

Borders on the Intersection of States, Technology, and Politics

At this time, borders of all kinds are emerging around the world in response to the COVID-19 pandemic.[1] SARS-CoV-2, the virus that causes this disease, has leaped across national boundaries, gripping populations and paralyzing social and economic activities. Where the virus appears, an unrelenting trail of victims follows. Countries, regions, cities, and villages are in lockdown, semilockdown, so-called intelligent lockdown, and less intelligent lockdown. People are in quarantine or isolation. While the virus seems to travel unrestrained, human mobility is governed by models, social distancing rules, and technical applications to prevent contamination. Borders are reappearing as new boundaries are spreading as rapidly as the virus itself. Although this book focuses primarily on borders in relation to human migration, I hope that its technopolitical and morphological analysis will also offer a valuable perspective for understanding the development of other kinds of borders and the mobility of other entities.

For anyone interested in the relationship among states, technology, and politics, and in questions concerning the inclusion and exclusion of persons, borders are almost inescapable. Coming from a philosophy of science and technology and a science and technology studies background, I became interested in the changing nature of borders around 2005. My interest intensified after a dramatic event that fueled my concerns and my curiosity. On the night of October 27, 2005, a fire broke out in the detention center in Amsterdam's Schiphol International Airport. Eleven detainees perished in the tragedy. Schiphol's detention center is at the airport's original location, in Schiphol East. When an individual arrives at the airport and is deemed not to have the right to stay, he or she will be held at this center as an irregular migrant, most likely awaiting repatriation.

This disaster and the subsequent investigation by the Dutch Safety Board triggered my interest in the tensions among globalization, technological development, and the ways that states select different sorts of persons. It appeared to me, and to many others, that a strange paradox had arisen in Europe: whereas more and more refined technologies such as iris scans at airports were being introduced to facilitate traveling and connect security to service, migrants were increasingly failing to reach their destinations, forced into taking risky routes or ending up in the shadow zones of Europe.

My investigation of the new technologies used in border control began shortly thereafter. Working for the Rathenau Institute, an organization that encourages public debate and advises the Dutch parliament on issues of science and technology, I analyzed the rise of technologies in migration policy: databases, fingerprinting, speech recognition technology, X-rays to determine the age of minors, deoxyribonucleic acid (DNA) tests to determine family relationships, and the like. In the following years, working at the Netherlands Scientific Council for Government Policy and the Department of Philosophy at the University of Amsterdam, my interest expanded to the politics of borders or, more precisely, the *technopolitics* of borders. Through my research projects, I became increasingly involved in various international networks of researchers.

Europe's borders are my point of departure to address contemporary geopolitical and humanitarian questions. Borders mark conflicts over international orders. Borders can turn places, routes, and territories into "zones of death." The seas between Africa and Europe have claimed thousands of fatalities—individuals who sought to reach a destination but were not considered regular travelers (e.g., because they were ineligible for a visa) and were thus barred from taking a safer route. But although borders are inextricably linked to all sorts of people on the move, it is not the people but the borders themselves that are the main characters in this book. My interest in Europe's borders is twofold. I explore the technopolitical dimension of borders, using particular approaches and concepts derived from science and technology studies and the philosophy of technology. I also use the analysis of borders to further explore and elaborate the concept of "technopolitics" itself.

The research that I undertook for this book coincided with an unprecedented episode in European border politics. Although developments concerning the borders of Europe, the application of technology, and the externalization of controls have been underway for some time, they came to a head during the so-called migrant crisis or refugee crisis—a period that continues but reached its peak in 2014–2016.

The Netherlands Scientific Council for Government Policy and the Department of Philosophy at the University of Amsterdam generously provided me with the time required to conduct the studies and fieldwork that form the basis of this book. I am extremely grateful to the editors of the MIT Infrastructures series, Geoffrey Bowker and Paul Edwards, and Justin Kehoe of the MIT Press for their enthusiasm, support, and guidance. The comments of anonymous reviewers were invaluable in improving the manuscript at the various stages. It is a huge privilege as an author to use all these comments to refine a book while developing my own voice.

A special thanks goes to Henk Wildschut, who gave generous permission to use a selection of his impressive photos on migration, camps, borders, and border controls in Europe for this book. His photos serve as the cover and return as chapter openers throughout the book. Wildschut's photographs of borders and migration are well known, and his work is exhibited around the world.[2] In January 2020, *The New Yorker* published his work on gardens that migrants make in refugee camps.[3] His projects on borders and migration resulted in 2010 in the book *Shelter* and the film *4.57 Minutes Back Home*, and other works followed, including the books *Ville de Calais* in 2017 and *Rooted* in 2019. His work has received prestigious awards, including the Dutch Doc Award in 2010, the Arles Prix du Livre in 2017 and the Netherlands Photo Book prize in 2019.[4] Without wanting to compare myself to his work, there is a certain similarity in the way in which his photos and this book depict borders. Like his work, the book pays attention to people and migrants of all sorts, but it is particularly interested in the various moves that borders themselves make. His work exactly represents the border situations that the book is about, without romanticizing or dramatizing them.

The time for me to spend researching this book was made possible by the Netherlands Scientific Council for Government Policy and the University of Amsterdam. An Open Society Foundations grant (OR2014–16667) supported my research on Greece and the United States. The fieldwork in Greece, partly in inspiring cooperation with Rogier van Reekum, would not have been possible without the support of Ermioni Frezouli, Vasiliki Makrygianni, and Aristotle Tympas. On Chios and Lesbos, Ermioni Frezouli helped to organize, conduct, transcribe, and translate my interviews; her work was invaluable.

My visiting scholarship at the Department of Communication at the University of California, San Diego, was cordially hosted by Kelly Gates and Chandra Mukerji. I owe a big thanks to Jill Marie Holslin and Norma Iglesias-Prieto for showing me around the US-Mexico border, and to Ricardo Dominguez for discussing the meaning of countersurveillance.

I also had the immense pleasure of working with Marieke de Goede, Jasper van der Kist, Annalisa Pelizza, Willem Schinkel, Lieke van der Veer, and William Walters in launching research projects, interpreting findings, and working on journal articles. At the Netherlands Scientific Council for Government Policy, I hugely enjoyed the inspiring conversations on Europe, migration, borders, and technologies that I had with Ernst Hirsch Ballin, Frans Brom, Emina Ćerimović, Corien Prins, and Mathieu Segers. With Dennis Broeders, I enjoyed many enriching and formative years that resulted in several joint publications. At the University of Amsterdam, it was a great pleasure to discuss borders, migration, security, and technologies with my colleagues Rocco Bellanova, Luiza Bialasiewicz, Robin Celikates, Beste Isleyen, Yolande Jansen, Barak Kalir, Beate Roessler, and Polly Pallister-Wilkins. Takeo David Hymans was of great support in editing the text.

Over the course of the project, as well as long before, it was wonderful to be able to share all my worries, progress and setbacks with my beloved life partner, Esther Miedema. Her question in the final stretch—"What did you actually want to say when you started writing the book?"—was the push I needed. Completing this book also means that I finally have to seriously consider the question of our boys, Manu and Luca (now twelve and nine years old), of whether I am finally going to write a children's book.

Across the research projects, paper presentations, lectures, workshops, writings, and coffee meetings, I benefited greatly from the comments, suggestions, interest, and support of numerous colleagues working on politics and technologies. I would like to thank Louise Amoore, Mark Brown, Vasilis Galis, Georgios Glouftsios, Charles Heller, Henk van Houtum, Bernd Kasparek, Koen Leurs, Debbie Lisle, Matthew Longo, Donna Mehos, Simon Noori, Jan Hendrik Passoth, Lorenzo Pezzani, Sandra Ponzanesi, Nicholas Rowland, Katerina Rozakou, Mark Salter, Kevin Smets, Thomas Spijkerboer, Jeroen Stout, Linnet Taylor, Peter-Paul Verbeek, Gerard de Vries, and Sally Wyatt. For their support in organizing several workshops and meetings,

I would like to thank the Department of Philosophy at the University of Amsterdam, the Amsterdam School for Cultural Analysis, the Amsterdam Centre for Globalization Studies, ACCESS EUROPE, and the Netherlands Graduate Research School of Science, Technology, and Modern Culture.

Plan of the Book

I hope that this book succeeds in navigating borders and their various concepts without falling into the trap of demarcating their boundaries too easily. Rather than closure, I aim to keep alive controversies about the political meanings and epistemological and ontological status of borders so as to be able to follow their movements. The book begins in chapter 1 by arguing that Europe's borders have become a different sort of entity. Rather than being demarcations of territory and expressions of sovereignty and jurisdiction, Europe's borders increasingly act as vehicles for politics. I begin with a concrete example of the mobile border and the tensions between migration routes, border controls, and Europe's migration policies in the mountain region between Italy and France. Through this discussion, I introduce the book's main themes, topics, and recurring concepts, most notably those of "mediation," "technopolitics," and "peramorphic politics." The notion of "peramorphic politics" derives from *peras*, the Greek word for "boundary," while *morphic* originates from *morphe* (meaning "shape" or "form"). The chapter ends with an explanation of the methodology and the book's focus on Europe, borders, and human mobility.

Chapter 2 brings this conceptual discussion to the borders of Europe. It offers a technopolitical account of Europe's borders since the mid-1980s and describes the development of the European Union's borders as the rise of an infrastructure. Borders as infrastructure have four crucial characteristics. First, border infrastructures connect large-scale networks with local situations and manifestations of borders. Second, border infrastructures select among migrants and other travelers in particular ways, not just by including or excluding them but by organizing specific forms of selection and circulation. Third, borders can be visible or invisible; there is a particular interplay between vision and action. Fourth, border infrastructures are often themselves movable entities. These characteristics all have organizing functions in the European Union's coordination of its internal and external borders. The chapter analyzes the emergence of particular border institutions, such as the Frontex agency, and the rise of border technologies such as the databases European Asylum Dactyloscopy Database (Eurodac), Schengen Information System (SIS) and Visa Information System (VIS), and follows the movements of the European Union's externalization of border control. Border infrastructures and their agents, institutions, and technologies appear to be laboratories that organize movement, arising out of all kinds of compromises between politics and technologies.

Chapter 3 delves deeper into the notion of the border as infrastructure by developing a morphological account of technopolitics. It reads the *Spheres* trilogy of Peter Sloterdijk through the lens of Bruno Latour's actor-network theory so as to engage with the morphology of politics. From this vantage point, borders are not only instruments of political decision-making, but vehicles of political thought and action. Borders do not just transport materialities and information systems; they also transfer political ideas and techniques. Although I question whether Latour and Sloterdijk have arrived at anything approaching a "political theory," their respective analyses are crucial for understanding the material form of politics and the way that politics, technologies, and materialities interact. Based on the ontology, spherology, and political theology of Latour and Sloterdijk, I further elaborate on the concept of peramorphic politics—a morphological technopolitical account of how borders and politics give and receive shape.

Next, to identify the morphological particularities of Europe's emerging border infrastructures, chapters 4, 5, and 6 analyze specific manifestations of Europe's borders as they bear on so-called mixed movements or mixed migration from outside Europe. Whereas chapter 2 gave an overview of the actors, institutions, and technologies that make up Europe's border infrastructures, chapters 4, 5, and 6 examine specific situations where these border infrastructures unfold.

Chapter 4 continues the examination of the morphological shape of technopolitics at Amsterdam's Schiphol International Airport. It studies the intermingling of design, detection, and detention at the airport by investigating their linkages and infrastructural compromises. Airports are focal points in the organization of human mobility, as well as places of restriction and selection that aim to manage and control international migration. These aims are pursued with all kinds of devices and architectural trajectories that enable smooth passenger flow, as well as security, border control, and migration management.

Chapters 5 and 6 examine the infrastructural compromises that emerged during the so-called migrant crisis of 2014–2016 on the Greek Aegean Islands of Chios and Lesbos. They address how actors, institutions, and technologies concerned with care and control combined, or failed to combine, border security and humanitarianism. I discuss processes of bordering from surveillance at sea to the migrant shelters on Lesbos and Chios. The two chapters, which contain detailed descriptions of the various practices of the mobile border, deepen current discussions about technopolitics in border, migration, and security studies. Chapter 5 engages with the European Surveillance Program (EUROSUR) and the European Union's hotspot approach, focusing on how situational awareness and interoperability result in a patchwork of instruments, organizations, and interventions. Chapter 6 describes the intermingling of the provision of care and control at the emerging humanitarian border and how its material components can be said to travel.

Chapter 7 turns to the infrastructural investigations of nongovernmental organizations (NGOs), volunteers, activists, and artists to open up the politics of the border. It describes their visual initiatives as well as media coverage, in the process advancing the notion of a visual cacophony—a "cacorama"—to characterize the visual politics of opening up border events. The discussion elaborates on technopolitics under the conditions of secrecy as migration and international security become increasingly entangled, with the movement of political thought and action via borders complicating accountability. Spatially and temporally dispersed border infrastructures are not clear-cut outcomes of political decision-making. Only with hindsight, after an event has occurred, will its consequences be open to public discussion.

Chapter 8 presents the conclusions of the book. It revisits the morphological notion of technopolitics as they bear on the four characteristics of border infrastructures identified in the preceding chapters. I return to the notion of peramorphic politics to describe the ongoing multiplication, transformation, and hybridization of borders and how Europe itself can be seen as a boundary project—as an infrastructural state with movable borders that organize mobility. In the final analysis, Europe's current relationship with borders renders borders—and Europe itself—as an extreme infrastructure obsessed with boundaries and limits.

The book ends with a coda on the relationship between borders and COVID-19.

Calais fence 1, July 2015.
Source: Henk Wildschut.

Calais fence 2, July 2015.
Source: Henk Wildschut.

Calais fence 3, August 2015.
Source: Henk Wildschut.

Calais fence 4, December 2015.
Source: Henk Wildschut.

1   The Border as a Vehicle

Migrating through the Mountains

Although the term "border" expresses delimitation and demarcation, it remains a concept with few limits. What was initially meant as a border, or which later becomes a border, can accrue functions not normally associated with borders. I state this not to preempt attempts to define "borders" in advance, but to emphasize that borders arise, change, perish, and continue in other guises.

Clavière is a municipality in the Susa Valley in Italy, at an altitude of 1,760 meters. The alpine village is home to a refuge known as Chez Jesus. "Chez Jesus" is not the original name of the building; nor was the building originally a refuge. Once a church, Chez Jesus is now occupied by a pro-migrant collective. On April 22, 2018, approximately 300 so-called irregular migrants and activists gathered at the refuge. After sharing lunch, the migrants and activists, some affiliated with the No Borders movement, began their "march against frontiers"—a 19-kilometer trek from Clavière to the French city of Brainçon—to protest the militarization of the border by state authorities and the local police.[1] The march was also a response to provocations by members of Generation Identitaire, the youth wing of Les Identitaires (formerly known as Bloc Identitaire), which was founded in 2012. Members of Generation Identitaire, patrolling the area around the Col de l'Echelle, used helicopters, drones, and jeeps to block the roads; destroyed signs along the path meant to aid migrants; and, acting as a militia, handed migrants over to the police.[2] In August 2019, three members of Generation Identitaire were fined and jailed by a French court.[3] By wearing uniform jackets, driving marked vehicles, and using "military language," they had led migrants to believe they were police officers.[4]

Why did the migrants want to cross the border in the Hautes-Alpes? To understand their route, we have to go to Ventimiglia—and to Europe's border policies in general. Ventimiglia is a seaside border town in northern Italy, four miles from the French Riviera. In the summer of 2015, the French government reintroduced border checks in an attempt to prevent irregular border crossings, using a provision in the Schengen Agreement of the European Union (EU) that allows states to introduce partial, temporary controls in case of emergency. But although the French government imposed ID checks on rail and road routes, it did not open centers where migrants could request asylum. Ventimiglia quickly became crowded with *transitanti* (migrants in transit); as many sought shelter in the town, camping under bridges, it was christened the "Calais of Italy" because the presence of migrants waiting for an opportunity to cross the border to France resembled the situation in Calais where migrants waited for a chance to reach the United Kingdom. In the meantime, the policies of the Italian government grew more repressive. Desperate attempts to reach France led to massive loss of life, as migrants drowned in the Roja River, were hit by cars, or were electrocuted on top of trains.[5] Commissioned by Médecins du Monde, the Dutch photographer Henk Wildschut chronicled the lives of the *transitanti* in Ventimiglia: the arrival of a young man from Nigeria at the train station, a man from Bangladesh walking back to Ventimiglia after being sent away from the border by French police, a man from Gambia looking toward the sea, envisioning a route toward a better future.[6]

Due to these containments, pushbacks, and forced transfers from Ventimiglia, migration routes—which had previously seen migrants walking through railroad tunnels or hiding in vehicles—moved inland into the Alps.[7] One region that saw intensified movement was the Roya Valley—dubbed the "vallée rebelle" by *Libération*—due to the aid given to migrants by local community members.[8] As thousands of migrants, including unaccompanied minors, sought to hike over the border, the French authorities again intensified their controls.[9]

Joshua F. left his home in Cameroon in 2016, traveling via Chad and Libya to Italy, where he reached the village of Clavière. During his journey, he fell victim to human trafficking, forced labor, and slavery. But unaccompanied children were often refused formal recognition as minors, and thus entry into France; assessments of their age often fell short of French regulations that require considering educational and psychological factors in a spirit of "neutrality and compassion." The practices of the French authorities did not only let down the migrant children. Although humanitarian assistance is protected under French and European law, the police intervened in search-and-rescue operations conducted by aid workers, volunteers, and activists.[10] Borders thus not only moved geographically, following the routes of migrants; they also penetrated local communities and societies. And when the passage through the Roya Valley in the Alpes-Maritimes became more treacherous, routes moved once again, this time to the Hautes-Alpes, including the Susa Valley and Clavière.

Irregular migration between Italy and France arises out of many tensions, such as between the migration policies of several European countries; the European Union's Dublin Regulation, which states that migrants should apply for asylum in the country where they arrive first; and the aspirations of many migrants to travel to countries in the north and west of Europe, where prospects are considered better. Migrants from Africa and Asia crossing the "frozen border" in the Alps; the control of the border by police and state authorities; humanitarian support from solidarity groups; and vigilante violence by members of anti-immigrant organizations—all illustrate the many meanings and movements of borders that I address in this book.[11]

Amid Moving Things

The borders of Europe run a gamut of policies, technologies, and interventions. The emergence of new migration routes, the rise of new borders and border controls, surveillance and patrolling by local police and state authorities, the rise of migrant solidarity groups and antimigration activism, renewed border controls in the Schengen Area—all point to tensions in and around the borders of Europe. What has befallen the once-almost-invisible borders of Europe? How can Europe combine open borders for a select group of countries, while maintaining border controls with others? How does Europe project itself geopolitically through its borders? What do these questions tell us about the very concept of borders? And what about the politics that accompany borders?

For some people, the fall of the Berlin Wall in 1989, the collapse of the Soviet Union, the drawing back of the Iron Curtain, and the reunification of East and West Germany heralded the coming of a borderless world. The expansion of the European Union, the expected spread of liberal democracies and open markets around the world, the introduction of open borders through the Schengen Area—all fueled an idea that Europe would leave behind the Cold War and prosper in a borderless world. But thirty years after the fall of the Berlin Wall, and despite the Schengen Agreement, Europe still has walls, permanent or temporary border checks between member-states, and a host of border control policies and technologies at its disposal to monitor and control the movement of people. All of this has become starkly visible since the so-called migrant crisis of 2014–2016.

My argument is not that Europe sleepwalked from a divided world into a borderless one, only to wake up in a world still divided by borders. My contention is that something else has changed—namely, the borders themselves. Since the Peace of Westphalia and the subsequent formation of nation-states, European borders—under the trinity of sovereignty, territory, and jurisdiction—have marked the authority of states and the boundaries of nations. But today, far from simple lines on a map, Europe's borders and border controls are vehicles for power, control, organization, coordination, and technology.[12]

Many of the issues that I address in this book can be seen through other lenses, such as "migration," "security," or "surveillance." But not all roads equally lead to Rome. Centering our analysis on borders allows us to focus on the relationships between states, international human mobility, and technology. The following chapters will trace the interactions of actors, institutions, and technologies that lead to the emergence of Europe's *border infrastructures*—a concept that in turn lends itself to studying the entanglement of digital, physical, and natural borders, various forms of governance, and the mechanisms of circulation, inclusion, and exclusion. The concept of border infrastructures also allows comparison with other kinds of information, communication, transportation, and security infrastructures as we study mobility and the conflicts and cooperation that it engenders. Border infrastructures shape new relationships between states and people, between initiatives to control and to care, and connect landscapes to seascapes, bringing geography back in.

In approaching borders, this book does not seek an omniscient view from high above mortal humanity—*kataskapos*, as the Greeks called it (from *kata*, "downward"; and *skopos*, "view," or "target").[13] My perspective rather begins *in medias res* (in the middle of things) in order to do justice to the granular and often haphazard ways in which borders infiltrate states and populations, landscapes and seascapes, and a spectrum of technologies and materialities. This chapter introduces three key concepts—technopolitics, mediation, and movement—that recur throughout the book; the concept of border infrastructures will be unpacked in chapter 2.

The concept of mediation will be fleshed out empirically in subsequent chapters, as we travel from borders on land and at sea to "smart" borders at the airport, from detention centers on the Aegean Islands to European data-bases and the headquarters of Frontex—the European Agency for the Man-agement of Operational Cooperation at the External Borders of the Member States of the European Union, in Warsaw. For now, note that the notion of mediation holds that the *interactions* among actors, institutions, and technol-ogies that constitute borders (in our case) generate *changes* that affect the con-stituting elements, from which new relationships and entities emerge. These changes take place via both the *transportation* and the *transformation* of the respective entities. Interactions are not restricted to social interactions; rather, they include the circulation of all kinds of information, knowledge, and tech-niques. Interactions do not leave the participating entities untouched; they affect the meaning and position of actors, institutions, and technologies, turning them into "mediators." Combining the functions of managing pas-senger flows, checking goods, and regulating migration at the airport leads to new connections and disconnections among the actors, institutions, and technologies that execute these tasks. When border security approaches to irregular migration interact with humanitarian approaches to provide medi-cal care and legal aid, notions of care and control become entangled, as do the respective actors and institutions.

One consequence of this view is that the border may appear omnipresent—that everything becomes a border or is perceived as such. We saw how bor-der controls and the actions of state officials, migrants, solidarity groups, and anti-immigrant activists in the Alps reproduced and transformed the border. Border controls do not take place only at international boundaries, but also inside and outside state territory at the hands of national border guards, as well as the personnel and techniques of private security firms. Any study of Europe's borders thus must resist turning into a reductionist "borderism." "Presentism" in the study of history means that the present is projected onto the past, whether in an explanatory or anachronistic way. "Borderism" analogously implies that all boundaries, policies that bear on mobility or circulation, and mechanisms of exclusion are seen as borders of some sort. What, then, would distinguish borders from national registries of citizens, urban surveillance practices, or identity checks to access certain services? Everything would become a border.

But there is also something to be said in favor of presentism as a way to study the coming into being of different presents. Scholars in science and technology studies have argued that the emergence of technologies cannot be explained by situating them in historical context. Technologies both change and are changed by society; for this reason, technologies should be considered sociotechnical constellations that create new associations and introduce new social, cultural, economic, and political relations and stratifications. Similarly, the emergence of borders cannot be explained only by situating them in their historical circumstances. Borders carry with them all kinds of political ideas, ideals, and motives; developed and deployed as instruments of authority, they shape, restrict, and broaden our repertoires of political action. For this reason, I conceive of borders as *mediators*.[14] Mediation emphasizes the relationship between technologies and societies and the construction of entities and events situated in between them. Events, furthermore, carry a certain presence: "mediation refers to the event, insofar as its possible justification by the terms between which it becomes situated comes after the event, but even more insofar as these terms themselves are then expressed, situated, and make history in a new sense."[15]

While the study of borders, migration, and state formation often focuses on international mobility, border control policies, and their humanitarian consequences, little work to date has explored the mediating work of borders. To do so, I work with concepts first developed in science and technology studies and in the philosophy of technology, and then elaborated in schools within political theory and international relations, as well as in parts of border studies, migration studies, mobility studies, critical security studies, and geography.[16] The concept of mediation focuses our attention on the fluid relations among borders, state authority, technology, and mobility. Mediations concern more than technical connection or political and administrative cooperation; they point to the emergence of novel forms of power, authority, and control, new sociotechnical arrangements that affect existing political orders. As the intensified border controls between Italy and France revealed, mediations include the transfer of knowledge and technologies, international cooperation between various practitioners, collaboration between state and nonstate agents, and the translation and implementation of political ideas and ideologies. Borders are themselves movable entities; they *shift* when state boundaries are redrawn and *transform* when new control mechanisms are deployed.[17] But the idea of borders as mediating entities points to something more fundamental.

The relation between motion, borders, and politics is a much-discussed theme in the critical scholarship on borders, captured in the concept of "kinopolitics"—"kino" derives from "kinetic" and the verb "kinein," meaning "to move."[18] A key idea in kinopolitics—the in-betweenness of borders— holds that borders not only reproduce the distinction between inside and outside, but also are motors that affect the deeper structuration of societies.[19] This accounting of borders underlines their mediating function and "spatial dynamic" as active entities that create centers and peripheries.[20] This mediating aspect of borders can be extended to the movement of all kinds of actors, institutions, and technologies.[21]

The intimate relationship between borders, materiality, and movement encourages us to take the notion of mediation a step further. The notion of *viapolitics* denotes movement and emphasizes the role played by vehicles— ranging from airplanes and boats to trucks and containers—in the process of migration and how they carry violence, struggle, and power.[22] Violent acts are not only committed by state authorities, as we saw in Generation Identitaire's use of helicopters and drones at the Col de l'Echelle. And as we saw in Ventimiglia, the "vias" include railroads and tunnels, the routes that migrants take, and their own bodies and legs as they cross forbidding mountain passes. But viapolitics does not apply only to vehicles and "things" that travel and the routes they take. Borders, I argue, can also be seen as vehicles.[23]

It may seem odd to relate borders and border politics to movement and circulation. Aren't borders about preventing particular people from entering? But as the example from the Alps and numerous studies have pointed out, blocking is but one function of the border. And although debates over international mobility are often couched in the binary of "open" and "closed" borders, borders are better seen as selection mechanisms that allow the circulation of *some* people, goods, finances, and information.[24] Walls that seek to prevent people from entering often fail in their aims, leading to "waterbed effects" and the search for alternative routes. During Europe's so-called migrant crisis, thousands of individuals, mainly from Syria, crossed the border between Russia and Norway at Storskog—on bicycles. They were exploiting a loophole in the rules: while Russia does not allow people to cross on foot and Norway does not let in drivers carrying passengers without documents, bicycles are permitted on both sides.[25] Walls can also lead to "now or never" moments that force migrants to take riskier routes or pay higher prices to reach their destinations. Borders may thus produce blockages in specific places while promoting circulation elsewhere, while walled states arguably project weakness rather than strength—a lack of economic and diplomatic means to influence human mobility.

Borders, in short, are mechanisms that organize circulation—of not only people but ideas and imaginations of international relationships among states and peoples. Whether it concerns the mobile provision of care and control in the traveling humanitarian border or the circulation of information and images enabled by sprawling databases, infrastructures of movement depend on all kinds of meticulous translations and associations, all sorts of small steps to bridge distances in time and space. These mediations in turn continuously lead to new infrastructures as the deployment of devices, instruments, and information systems expands.[26] Emphasizing the "kino" and the "via" and the variety of objects and subjects that move across and through borders paves the way to further conceptualizing the mediating role of borders.[27]

Technologies and Borders

Europe's borders are focal points where questions about states, politics, technology, citizenship, and international mobility come together. Borders give rise to questions about people's rights, inclusion and exclusion, state sovereignty, authority, jurisdiction, identity, and belonging. Borders come in many shapes, including checkpoints on land, at sea, and in airports; they encompass passports and travel documents, visas and databases, asylum procedures, and deportation policies. Borders are variegated entities that operate in different ways across space and time, affecting almost everyone who travels—citizens, expats, tourists, and migrants—albeit in starkly different ways. Borders can induce racialized and gendered violence and pose specific risks to specific groups of people aspiring to mobility. Border controls in Europe also affect sedentary populations, as the regulations, registrations, and techniques governing international mobility require ever more data and secure identity checks in an interconnected world. History is replete with policies developed to address specific problems which were later applied more generally. The fingerprinting policies of the United Kingdom (UK) were not developed by Scotland Yard—they arose in the colonies, particularly India, in the second half of the nineteenth century as colonial administrators sought to tackle the problem of fraud and impersonation. What began as handprinting developed into fingerprinting and a technique for civil identification before it became a tool for identifying criminals.[28]

Borders have a particular relationship with technology. The term technology not only alludes to the technical aspects of devices—tools, materials, machines, instruments, and computer networks—but how they are designed, produced, and managed. Technologies inform—and limit—how societies are governed and can be imagined to be governed. Like technologies, borders are both concrete and abstract, both material and ideational.

Take, for instance, how borders affect time. Borders can delay or accelerate the crossing of boundaries. For some travelers, automated borders at airports speed up checking in and passport control; for others, borders mean waiting for papers and queuing at embassies. Databases function as archives of human mobility, storing information about travelers in order to recognize patterns and make decisions in real time. Border-monitoring technologies aim to arrive at *situational awareness* by connecting interventional space to real-time imaginaries. Social media affects the organization of time as migrants travel through places and spaces, while online mapping platforms allow people to report critical situations where migrants are in danger. Borders can unite the histories of peoples and nations; recall colonial or violent pasts; and encourage dreams of reaching destinations and the future. In all such cases, technologies inform existential and political questions; and devices, networks, and infrastructures point to a material locus, as well as a particular form of politics.[29] Technological transformations, moreover, produce a variety of temporalities. Introducing new border technologies will favor some groups of migrants while disadvantaging others, such as those granted visas and able to enter countries by plane versus those who have to take other routes to apply for asylum. Migration is mediated by all kinds of information and communication technologies.

Technology also guides the political imagination. The border between the United Kingdom and the Republic of Ireland in the wake of Brexit has once again become the subject of heated debate over "hard borders," "soft borders," "visible borders," and "invisible borders." While any visible hard border between the Republic of Ireland and Northern Ireland recalls the violence of "the Troubles," the transition that began on January 31, 2020, envisions a new model to regulate the movement of people, goods, capital, services, and information. All kinds of technological solutions are now being considered. Options include "trusted traveler" and "trusted trader" schemes, digital platforms for the preclearance of travelers, automatically processed customs systems, the electronic identification of vehicles through number plate recognition, and the certification and global positioning system (GPS) tracking of goods. But regardless of how virtual or invisible they may appear, all these technologies come with registration, administration, checks, risk analyses, and surveillance systems. As an expert on the Irish border stated in the *Irish Times*, "a 'frictionless' border is almost an oxymoron."[30]

To clarify the relationship between technologies and borders, I critically elaborate on the distinction between an *ontic* view, which sees technologies as instruments or as outcomes of human design, and an *ontological* view, which suggests a more intimate and co-constitutive entanglement between humans and technology.[31] In the ontic view, technology is usually defined in instrumental or anthropological terms—the former sees technology as a means-ends relationship, the latter as the result of human activity. In the ontological view, technology harbors a mode or attitude toward reality that unveils something about the relationship between humans and technologies, and ultimately about our "being."

This distinction can be illustrated through the technologies of border control. A wall to prevent irregular border crossings, even if ineffective or counterproductive, is a means to an end. Other examples of the ontic view include legal approaches that seek to separate technologies from their uses and meanings, such as when the consequences of a database are evaluated using the purpose limitation principle that restricts the collection, processing, and use of data to specific purposes to prevent function and data creep. But deployments of technology raise other issues. The use of Big Data for security purposes creates new categories of travelers considered a risk; facial recognition technologies can lead to profiling and discrimination; and speech recognition technology can be used to test the language proficiency of people seeking residence, although experts disagree whether such software is sufficiently accurate for this purpose. Technologies also affect the discretionary powers of civil servants. Personnel who question asylum seekers use decision-supporting tools such as maps with detailed regional information and automated decision trees to check countries of origin; such tools not only support but prestructure decisions so that it is hard to distinguish which parts of the decision to grant asylum were driven by technology. When technology is conceived in relationship to being, an even broader picture emerges. Technologies to control borders are no longer evaluated by their particular uses; rather, they are part of an emergent machinery that informs our thinking about territory, sovereignty, and human mobility.

However, between *ontic* and *ontological* views, many more forms of borders, technology, and politics are available. The study of borders and technologies widens our conceptual repertoire to analyze the relationship between technologies and politics, showing that there are many more options in between reducing technologies to instruments and enlarging them into expressions of our relationship to being. The politics of technology can be conceived in myriad ways, from technology being essentially political, to being a form of politics by other means, to being the site of political contestation, to being a part of our collective material "world making."[32] What remains to be investigated are the analytical schemes that underlie these various conceptions of politics and technology—to unravel how they relate to and may lead to each other. Some repertoires see borders as objects or instruments of state power; others see borders as networks, as large-scale configurations that organize international human mobility; yet others see borders as a kind of worldview, a way of ordering reality. Next, I present a technopolitical account of borders that privileges the *transformations* between these repertoires by following how technologies travel from one form to another.

The Technopolitics of Borders

Emphasizing that technology, in the context of migration and border control, has a political dimension may seem like stating the obvious. Why does the study of borders, migration, surveillance, and security require this extra emphasis on political issues and concepts? Few people will deny that policies surrounding international mobility and migration stem from political choices regarding citizenship and rights, inclusion and exclusion. The technologies involved *always* aim to control, surveil, monitor, and select as they serve political goals, or because they are deployed in highly politicized situations where risks are assessed and decisions about visas or asylum must be made. Nevertheless, there remains something peculiar about iris scans, fingerprints, databases, detention centers, radar images, patrol boats, and barbwire fences. There are important differences in claiming that things are designed to serve a political goal—that technologies are part of a political constellation—and that things generate political consequences. Although something political is at stake in all of these cases, we need to examine different conceptions of technopolitics to clarify what they mean by "politics."

To begin with, definitions of technopolitics need not consist of a combination of a definition of "technology" and a definition of "politics." The literature on borders, migration, technology, and security has produced a plethora of concepts to distinguish among particular kinds of technopolitics.[33] The constitutive and sometimes decisive role of technologies and materialities is widely acknowledged, empirically as well as conceptually.[34] For example, theorists inspired by the work of Michel Foucault generally understand technology in the broad sense of the term, referring not only to concrete things, networks, and instruments but to the arrangement of techniques that govern subjects, the relationships between technology, power, and knowledge, and how subjects and objects are created by particular ways of seeing and knowing. Notions of "governmentality" and biopolitics and how policy apparatuses govern populations and territories have informed analyses of border regimes, security policies, and state regulations; refugees and humanitarian governance; citizenship; and international mobility regimes. These notions also encourage us to see policies as a kind of technology organizing the governance of borders and the movement of people.

Analyses in science and technology studies, most notably elaborating on Bruno Latour's work, to which I will return in detail in chapter 3, show that technologies are not only neutral means to achieve certain goals, as they transform goals and needs. While the presence of technologies has made long-distance communication and curing diseases easier and more efficient, it has also altered how we communicate and what we treat as a disease. Technology is thus seen as a repertoire of actions, interventions, and representations through which issues can be articulated—a conception that allows us to rethink the sociotechnical relationship between humans and technology. As both instruments and expressions of being, technologies are indispensable linkages in the sociomaterial world.

Despite the wide attention that is already being given to the relationship between technology and politics, there are some specific characteristics that deserve further conceptual examination, particularly where borders are concerned. The morphological conception of technopolitics that I develop in this book foregrounds three aspects—*materiality, mediation,* and *movement.* So what does a morphological approach to technopolitics entail? Morphology is the study of shapes. Rather than providing an analysis of various kinds of technopolitics or taking the study of policies, regulations, and border regimes as a starting point, a morphological approach aims to identify *forms* of technopolitics by analyzing how politics and technologies intermingle. My morphological approach does not a priori view technology as an instrument, network, or worldview; it holds that technopolitics can take many shapes and transform from one to the other. It focuses on the transformation between repertoires that consider borders as objects or instruments, borders as networks, and borders as a kind of worldview.

In contrast, normative philosophical approaches often view borders as the objects of policies, instruments of legislation, obstacles for migrants, or the avatars of politics. These perspectives are justified, both conceptually and empirically. In practice, borders are often understood and applied as instruments. Understanding technology as a means to an end fits well within existing institutional politics, as it allows distinguishing between goals and implementation in order to better organize the spatial, temporal, and functional division of tasks and responsibilities. Nevertheless, instrumental approaches often ignore how technologies develop and change, the gradual merging of their goals and functions, and their unforeseen or underestimated side effects. Normative approaches distinguish between humans and nonhumans, focus on decision-making, and hold that technologies can be designed and governed according to political will. But such approaches hit their limits when it turns out that objects cannot be easily isolated from other elements in the surrounding environment, and when their becoming political brings about a collective transformation of related entities. Distinctions between subjects and objects are even harder to maintain once we see that decision-making, the anticipation of risks, policymaking, and political thinking are all suffused with technology. The understanding of borders as technopolitical entities requires a broader perspective in order to take these meanings into account.

Nevertheless, the focus on technopolitics is far from self-evident. The first risk is that emphasizing the networked nature of borders complicates demarcation: how do we decide which situations and forms of violence, exclusion, and oppression can still be said to relate to borders, and which ones not? Network analyses have been criticized on the grounds that it is often impossible to define a network's boundaries—a criticism that surely resonates if borders are deemed to be everywhere. Where, then, does the border begin and end? But network studies and situational attention need not be mutually exclusive. "Networks can take any scale—have the power to cross different organizational levels—precisely because each relation invokes a field of embodied [social] knowledge about relationships."[35] If the border is everywhere, it can be analyzed everywhere in its different manifestations. The form of organization required to connect the various settings is arguably more important than the breadth of the network or the scope or scale of the infrastructure.

The second risk of (particularly my morphological understanding of) technopolitics is that focusing on the multiplication of borders and the technological machinery of bordering renders travelers and migrants and the injustices they endure invisible. Celebrating hybrid formations can obscure enduring asymmetries between actors and agents.[36] Opening up configurations of agents, institutions, and technologies in order to study processes of inclusion and exclusion, the creation of new asymmetries and forms of violence is thus necessary to keep the political nature of networks and infrastructures in plain sight. Analytically privileging the workings of technology should not mean that international political developments and situational particularities—the cultural, economic, and historical dimensions of the places where technologies operate—fade from view. While focusing on technopolitics emphasizes the role of all kinds of nonhuman entities and devices that make up the border, we still need to attend to violence, discrimination, oppression, and exclusion.

The morphological account of technopolitics presented in this book will focus on the development and dissemination of political ideas via their manifestations in all kinds of technologies and how those ideas are affected and change. Border control technologies such as walls, fences, cameras, and databases are not only instruments that can be acted upon at will; the following chapters will explore the idea that border politics develop *through* technologies as well.[37] Technology is not just an instrument, a device like a hammer that we use for a specific purpose. Nor is technology just an extension of the body, a prosthesis. Yes, glasses improve our vision, but something else is involved too. The morphological view holds that knowledge and ideas are realized and unfold through the development of technologies.

To develop my morphological account of the technopolitics of borders, I introduce the notions of *peramorphic mediation* and the resulting *peramorphic politics*. *Peras* is the Greek word for "boundary," "limit," or "end"; and *morphic* originates from the Greek *morphe*, meaning "shape" or "form." The term *peramorphic* can be used descriptively, but also pejoratively. "Anthropomorphic" means having the form of a human, while "anthropomorphism" is often used pejoratively, implying that human characteristics are misleadingly attributed to something that is not human. In a similar vein, the term *peramorphic* can be used to describe something turning into a border, but also to argue that something is unfairly or mistakenly seen as a border. Once a border is created, more are likely to follow—in different shapes at different locations. Borders are contagious; processes of mediation always create more and new borders.

My notion of peramorphic mediation, to be discussed in detail in chapter 3, focuses on the creation, reproduction, and distribution of borders by all kinds of translation processes and the intermingling of borders with movements and materialities of all sorts. The resulting morphological understanding of the technopolitics of borders—peramorphic politics—acknowledges the intrinsic tensions contained within borders: between closing and opening, between demarcating a certain space while opening or defining another. In the first instance, peramorphic politics can be seen as a *limited* kind of politics. It sets its own boundaries by pursuing a limit—narrowing the repertoire of technopolitics in which compromises must be made between circulation and isolation, mobility and fixity—into a program oriented at an *eschaton*, a final event or last thing.

Then again, peramorphic politics is more concerned with the opening that follows the ending, the horizon that offers a window on future events. A border is not only something that blocks; it is also "that from which something begins its precensing."[38] A border, in this view, is a kind of horizon. What glows on the horizon is space, created by a boundary. But borders in this line of thinking still lack shape and volume, and they are pictured as mere lines or perimeters—a flat horizon. Whereas boundaries may be the starting point of something else, borders themselves fall out of view as entities. For this reason, I develop over the course of the following chapters a peramorphic approach that acknowledges borders as entities with a particular extensiveness.

Following Border Infrastructures

Scholars in science and technology studies often investigate the emergence of things by *following* the actors that constitute them—thus studying things in the process of becoming or, in our case, infrastructures in the making.[39] Methodologically, the "following" concerns human actors, social and political institutions, and technologies, while focusing on their interactions allows for a perspective that acknowledges that political ideas and policy practices often co-emerge.[40] While political decisions may enforce certain policies and programs, their exact meaning often only emerges in their making. An example that will recur in the coming chapters is the development of EU databases for visas, asylum, and migration, such as the Visa Information System (VIS), the Schengen Information System (SIS), and the European Asylum Dactyloscopy Database (Eurodac). The meaning of the phrase "digital border surveillance" unfolds through the interoperability of the various databases and the interactions of the actors and institutions using them. The meaning of the concept relies on the consequences it brings through practice.

Applying the methodological and conceptual notion of "symmetry" to borders, states, and migrants raises thorny questions. Broadly speaking, the literature on borders and migration harbors a tension between views that privilege the "tactics of bordering" and the "autonomy of migration," respectively—a palpable tension when these are presented as key poles in the politics of Europe's borders.[41] The autonomy of migration argument begins with the premise that human mobility is a right undermined by regulations, obstructions, and interventions, by documents, walls, visas, and violence. From this perspective, borders are never natural entities but constructions that constrain human mobility. Similarly, the notion of "migration" itself is seen as a particular way to conceive of human mobility, namely as movement in a world governed by states, jurisdictions, and borders regimes. The risk of designating human mobility as "migration" is that it naturalizes the border.[42] Borders are indeed material compositions that can be seen as expressions of authority and jurisdiction, as manifestations of state power and the governance of human mobility—as vehicles of politics, thought, and action. But neither is human mobility—fueled and mediated by a range of social, economic, ideational, and technological factors—a natural phenomenon. Rather than holding that "if there were no borders, there would be no migration—only mobility,"[43] I suggest that if there were no state borders, other kinds of borders would emerge, in different shapes, at other locations.

The vignettes at the beginning of this chapter revealed the mobility of migrants, borders, and border controls; the We Are Here movement further illustrates how borders keep moving with migrants. This movement mainly consists of undocumented migrants who collide with borders wherever they go. The movement began in 1997 with the so-called Caravan of Migrants in Germany; the "We Are Here" slogan has its roots in the United Kingdom, where black and South Asian antiracist activists in the 1980s chanted, "We are here because you were there." In the United States in the 1990s and 2000s, undocumented migrants chanted "*Aquí Estamos, y No Nos Vamos!*" (Here we are, and we're not leaving!), often accompanied by "*Y Si Nos Sacan, Nos Regresamos!*" (And if they throw us out, we'll come right back!).[44] The city of Amsterdam has also seen manifestations of this movement, with undocumented migrants gathering at a bunker in the Vondelpark and in empty churches, schools, and office buildings.[45] The intriguing aspect of these actions is not only that undocumented migrants claim a position in the public sphere, but also that their very actions create a public sphere, which—like the border—they carry with them.[46]

Two lessons can be drawn from all this. First, we need a symmetrical perspective to simultaneously examine human agency and technological structure—one able to transcend the dichotomy between the subjects and objects of border control technologies (i.e., the distinction between those who develop the infrastructure and those who are subjected to it).[47] In other words, borders move not only with the steps that states take to reposition them, but with the movements of migrants and their material means of movement.[48] Second, borders are not always the robust and durable building blocks they proclaim to be. In border surveillance and mobility management, they are often patchworks, outcomes of diplomacy under highly politicized circumstances and the cut-and-paste work that comes with creating "systems of systems" that aim at "interoperability," giving rise to various forms of "data friction."[49] The infrastructures that monitor mobile agents are themselves mobile.

The conceptual and the empirical thus travel and change together.[50] As we saw in Ventimiglia, the technological construction of the border is not the simple result of a clear-cut decision-making process translated into policies and executed by state officials and private contractors; the actual border resembles a patchwork where conflicting desires, existing patterns, and new trends are woven together. Conceptually, there are two important points. First, a multiplication of labor takes place via all kinds of border practices and activities. Second, a great deal of translation work is required to relate concepts such as "labor" to concrete material situations.[51]

My focus on materiality thus has two aspects. The first is to acknowledge the role of material constructions such as databases, detention centers, and rescue operations as sites of border politics, as well as the role played by material devices ranging from migrants' cell phones to the design of airports within border infrastructures, which will open up practices, technologies, and relationships to examination. The second is to acknowledge that accounting for nonhuman elements requires a different approach. Respecting the transformative role of materialities requires a conceptual repertoire able to detect the interactions and translations taking place between actors, institutions, and technologies. These repertoires must also be able to follow transformations as technologies travel and intermingle with landscapes and seascapes. Following interactions between humans and technologies will yield unexpected encounters, as border technologies not only consist of high-tech information systems, but entail all kinds of mundane materialities as well.

The Boundaries of Europe

While I focus on Europe and Europe's borders, many of the developments described in this book are driven by the decisions and policies of the European Union. It is important to note that Europe and the European Union are not synonymous. When I refer to EU member-states, institutions, technologies, agencies, regulations, policies, and politics, the reference will be obvious. But the control of Europe's borders is not limited to EU policies and politics. The European Union involves nonmember-states in its border politics through partnerships and associations. Cyprus, Ireland, Croatia, and Romania are members of the European Union but not part of the Schengen Area; meanwhile, Lichtenstein, Iceland, Norway, and Switzerland are not members of the European Union but participate in Schengen. New countries may join the European Union; some are already candidates, including Albania, the Republic of North Macedonia, Montenegro, Serbia, and Turkey; others are potential candidates, such as Bosnia and Herzegovina. And of course, the United Kingdom has left the union. EU policies also affect border control in countries in Africa. Partnerships with African countries and the so-called externalization of border control transfers EU border politics across borders.

Chapter 2, which recounts the rise of Europe's border infrastructures, begins in the mid-1990s but focuses in particular on the years 2014–2016, often referred to as the years of the "migrant crisis" or "refugee crisis." I reproduce these terms with the proviso that this so-called crisis was actually a multiplicity of crises: a political and policy crisis of decision-making, a crisis of solidarity, and above all, a humanitarian crisis. Although this period was exceptional in many ways, the events also revealed a great deal of continuity from earlier policies, plans, geopolitical developments, and humanitarian dramas. Since 1960, the number of migrants as a percentage of the world's population has remained remarkably stable, at roughly 3 percent. The idea of a migrant or refugee crisis also has to be understood in context, as refugees represent only between 7 and 8 percent of the global migrant population.[52] The terms "crisis" and "exceptional" also run the risk of—intentionally or unintentionally—legitimizing certain actions and responses to avert an alleged crisis or to justify exceptional measures to solve it. While I do not avoid these terms, they need to be used with caution.

The distinction between refugees, migrants, and asylum seekers deserves attention. This book is not about migration within the European Union (largely consisting of labor migration within the single market), but about migration to the European Union from so-called third countries—countries that are either not part of the union or countries and territories whose citizens do not enjoy the right to free movement that the union enjoys. But this still covers many forms of human mobility. More precisely, this book focuses on the relationship between borders and what are called "mixed movements" or "mixed migration" from outside Europe. According to the United Nations High Commissioner for Refugees (UNHCR) and the International Organization for Migration (IOM), the term "mixed movements" refers to "flows of people traveling together, generally in an irregular manner, over the same routes and using the same means of transport, but for different reasons. The men, women and children traveling in this manner often have either been forced from their homes by armed conflict or persecution, or are on the move in search of a better life."[53] The European Commission defines "mixed migration flow" as a "complex migratory population movement including refugees, asylum-seekers, economic migrants and other types of migrants as opposed to migratory population movements that consist entirely of one category of migrants."[54]

Although the migration movements that I address in this book include asylum seekers and refugees, in the following chapters I generally use "migrant"—an umbrella term that includes refugees and asylum seekers. Although many people who arrive in the European Union seeking international protection can be considered refugees in a humanitarian sense, from a legal point of view it would be inaccurate to call them "refugees" until they are legally granted refugee status. But there is another reason to apply the umbrella term "migrant," and to refrain from a hasty categorization in migrants, refugees, and asylum seekers. Doing so would introduce a false dichotomy between people who, after long procedures, are considered refugees in the legal sense of the word and people who are not considered refugees but are nonetheless looking for refuge for good reasons. Creating such categories is one of the functions of selection processes at the border. We need to exercise due caution when categorizing migrants.

Although Europe's current borders and politics are far from sui generis, they have some unique features. First, the European Union is not a nation-state but a union of states; this affects the role of state apparatuses in border control and how authority, sovereignty, and jurisdiction over territory and mobility are expressed. Second, Europe's borders are highly variegated, consisting of land, sea, and air borders, internal borders between member-states, external borders with neighboring countries, and surveillance efforts inside and outside European territory. Third, Europe's borders have seen dramatic changes over the past decades. The expansion of the European

Union with new member-states, the reunification of the two Germanys, the creation of the Schengen Area, and the various treaties, agreements, and partnerships on migration and border management with countries outside of the European Union have turned Europe into a transnational border project.[55] These particularities notwithstanding, the highly dispersed and moving borders of contemporary Europe can be seen as places where issues concerning territories, infrastructures, authority, technology, and mobility interact. Findings from studies elsewhere that emphasize the disseminated nature of borders apply to the European situation as well.

The European Union's geopolitical and security interests provide a window for studying the entanglement of politics and technology, infrastructures, and states. As security, migration, and foreign policies become increasingly linked, borders are the entities par excellence to use to study the composition of power and infrastructure and the frictions and the political and technological mediations they bring about. Much more than the instruments of states, borders bring about all kinds of interactions among countries, state authorities, private firms, travelers, and technologies. The issuing of passports, visas, identity cards, and travel documents is intimately connected to the installing of national and international registrations, databases, checkpoints, and monitoring instruments. The threat of terrorism in Europe following 9/11 and the attacks in Madrid in 2004 and London in 2005 entangled border control, migration, and security policies and strengthened the external dimension of Europe's border politics, which now consist of an extensive infrastructure of identity checks, control mechanisms, surveillance instruments, and security policies that intervene inside and outside Europe. Borders are not just lines on a map or stripes through the landscape; they have volume, spatially and materially.

As a union of states, the European Union as a political entity differs from "states," "federations," and "empires."[56] But its border politics have many similarities with those of states around the world. Twenty-first-century borders provide vignettes of the intensified efforts to control international human mobility, of how sovereign authority and control over territory are currently expressed. While the boundaries between the United States and Mexico and the barriers and surveillance systems deployed by Israel in the West Bank are considered iconic borders, many more examples worldwide showcase similar border politics and technologies. Whether it is Australia's ocean patrolling, instituting biometric travel authorization in Uganda, or surveilling the Bengal borderlands, technologies, geographies, and authorities are intimately entangled at the border. Europe is no exception. Fences surrounding the Spanish enclaves of Melilla and Ceuta in Morocco, registration centers in Lampedusa and Sicily, barbed wire in the Balkans: local manifestations of the border may be said to be everywhere.

In the meantime, borders have been stretched, displaced, and transformed. Automated border controls at airports, the sanctions against private carriers transporting undocumented travelers, and the mobility partnerships between the European Union and third countries belong as much to the borderscape as refugee camps and detention centers. European countries have also copied ideas and practices from other places. The European Union's planned entry-exit systems are inspired by the Biometric Identity Management system of the United States, formerly known as US-VISIT. The European Border Surveillance System, known as EUROSUR, follows in the footsteps of previous surveillance programs in Spain and North Africa, while Israeli security systems and border surveillance technologies have influenced EU policies. A further factor is the growing role of the transnational security industry, with private companies promoting the integration of various systems to gather, interpret, use, and transport data and information for multiple applications.[57]

Breaking with European exceptionalism is not to suggest that a global border regime has emerged. Although border configurations and border politics may have many things in common around the world, border infrastructures always manifest themselves under particular circumstances in place and time and harbor specific selection mechanisms. The different visa regimes used by European countries and the particular selections and distinctions they draw—between migrants from "Western" and "non-Western" countries, former colonies and countries without a shared colonial past, or labor and knowledge migrants—often produce arbitrary and/or discriminatory categorizations that lead to unequal, racialized, and gendered forms of inclusion, exclusion, and circulation.[58]

These specific selection processes underline once more that borders are saturated with tensions and conflicts, answer particular needs and questions, cause specific problems and are informed by particular forms of knowledge, expertise, and technology. Rather than being the expression of fixed states or nations, borders transform political entities, while border infrastructures fuel identities and imaginations as states and nations. Studying border infrastructures is therefore also a means to study broader political transformations. As such, the other chapters in this book aim to not only offer a novel conceptualization of technopolitics and a tech- nopolitical interpretation of borders, but a description of Europe and the project of the European Union, as driven by the development of border infrastructures.

Watchtower, Bulgaria, September 2015.
Source: Henk Wildschut.

2   The Rise of Europe's Border Infrastructures

Europe's Borders as a Circulation System

If borders are technopolitical entities to organize circulation, what kind of framework is best suited to study this circulation? How best to describe the relations among different actors, institutions, and technologies, ranging from quotidian tools to large-scale information networks? Rather than characterizing Europe's borders, border controls, and migration management as a kind of fortress, panopticon, or form of surveillance, I opt for a perspective that can accommodate myriad policies and practices leading to different technopolitical configurations.[1] For this reason, I frame Europe's technological border networks as a kind of infrastructure.[2] Like other infrastructures, borders consist of construction works, communication networks, coordination centers, monitoring instruments, and networks of employees, officers, and technical experts. Border infrastructures are also connected to many other infrastructures, including those of data, transportation, knowledge, security, finance, and humanitarian intervention. The movements and materialities of these infrastructures are intimately tied to the formation of states and, in the case of the European Union (EU), a union of states. Border infrastructures provide a particular way to understand Europe. Rather than viewing European cooperation and integration—and sometimes disintegration—in functionalist terms or as a grand design, border infrastructures can be described from the inside out by following how they organize preventions, selections, and interventions. This chapter begins our investigation by tracing the recent development of Europe's border infrastructures and by unpacking the relationships among agents, technologies, and institutions.

Europe's borders span jurisdictions and technologies and are vehicles for political thought and action. As historians of technology have shown, processes of cooperation and integration in Europe were initiated by experts and technical communities, supported by technological advances and infrastructures of all kinds.[3] Whether they concern canals, railroads, radio broadcasting, or electricity grids, studies of technology shed a refreshing light on the things that hold societies together, as well as on dynamics within international relations.[4] Opening up these technologies and their histories often leads to nuanced understandings that temper all-too-general theories of functionalism or grand teleological schemes. Unraveling the technological "tensions of Europe" requires the study of all kinds of situated knowledge, technical possibilities and impossibilities, and, perhaps most of all, the often-contradictory political motives behind them.[5]

The study of borders and technology parallels the study of other material infrastructures.[6] Roads, railroads, water pipes, and electricity grids, to name a few, have been central to the formation of modern nation-states, in establishing state control over territory, and in the creation of the European Union.[7] Infrastructures are never tension free, as they select and create so-called winners and losers.[8] Border infrastructures are notable in that they explicitly seek to make selections—to distinguish between persons allowed and denied entry.[9] Four characteristics will help us to better understand Europe's borders *as infrastructure*.

First, borders-as-infrastructure consist of large-scale networks that connect to particular local situations. Not all infrastructures are grand projects designed and implemented from above; they also emerge out of singular events that form the building blocks of later structures.[10] The notion of infrastructure does not reduce myriad technological policies and practices to a single constellation. For example, border control infrastructures on land, at sea, and in the air (airports) have crucial differences; borders can also intermingle with nature and render different terrains into borders or stand out materially from their surroundings. Rather than distinguishing between borders on the basis of their natural or physical features, I focus on the traveling of methods and materialities applied in different situations.[11] Infrastructures shape common worlds, not by directly providing public goods or shared facilities but by fabricating particular connections that shape all kinds of associations between people and technologies. Borders-as-infrastructure are composed of myriad linkages between states and people, public and private, connectivity and collectivity.[12]

Second, although all infrastructures privilege some people over others—such as through variegated pricing, differential access to users, or quality of services—one of the main tasks of border infrastructures is to intentionally *exclude* at least some groups of people. While borders express a certain idea of belonging, membership, and citizenship for some, they deny this status to others. The border controls of the European Union can be seen as part of a highly political integration process—one with strict selection and prevention mechanisms and severe humanitarian consequences. The staggering number of people who have lost their lives on their way to the continent of Europe is a grim indication of how the management of mobility and the selection of membership work out in practice—and how the governance of international migration can intensify the differences in the life chances and expectations of geographically separated peoples. As such, European migration, asylum, and border management policies are in line with the international trend of states redefining their policies to privilege specific groups of migrants and asylum seekers while complicating access to others.[13]

Third, borders-as-infrastructure enter the realm of the visual. Infrastructures are related in particular ways to the interplay of the visible and the invisible.[14] What border technologies have in common with other infrastructures is that patchworks, or systems of systems, appear as seamless webs. But this image obscures the tensions and ruptures that any infrastructure must overcome to appear as a unified whole. Infrastructures are composed entities that visualize and disclose specific events at various moments in time and space.[15] This applies to the European Border Surveillance System (EUROSUR), which aims to link diverse national monitoring and registration systems by way of interoperability. It also applies to Europe's reception and detention centers. On the one hand, these closed camps function as the "black holes" of the migration regime. On the other hand, they are subjected to daily media coverage and seem to function as frightening examples to deter future migrants.

Fourth, borders are movable infrastructures. "It is not [only] migrants who migrate, but rather constellations consisting of migrants and non-migrants, of human and non-human actors."[16] Borders move not only with the steps states take to displace them, but with the movements of migrants and their material means of movement.[17] For this reason, I employ a symmetrical perspective that transcends the dichotomy between the subjects and objects of border infrastructures—those who develop the infrastructure and those who are subjected to it. State representatives, international organizations, and nongovernmental organizations (NGOs) move with migrants, implementing migration and security policies and providing humanitarian care. Migrants are simultaneously subjects and objects: their mobility makes them the objects of state attention. The border, then, is not a passive entity waiting to be crossed. It is a movable entity; border infrastructures move with people.

The history of state formation, as well as the history of the European Union, are about bordering and rebordering, of struggles and tensions, as well as of unification and belonging. In these histories, borders are not static, clear-cut demarcations between regions and states; rather, they are movable objects that can be shifted to intervene where the movement is. Many of Europe's border control technologies—such as databanks, information systems, registration centers, and monitoring tools—are not deployed at the physical border but rather organize mobility from a distance. Borders themselves can be highly mobile devices that pop up where the action is, sometimes even traveling alongside migrants on the move. Borders, movement, and materiality are intimately related.

This chapter analyzes the emergence of Europe's border infrastructures by following interactions between actors, institutions, and technologies. I begin by recounting the development of the Schengen and Dublin systems, the emergence of Frontex, and databases such as the Schengen Information System (SIS), the Visa Information System (VIS), and the European Asylum Dactyloscopy Database (Eurodac). Particular attention will be paid to border control efforts during the so-called migrant crisis of 2014–2016: the EUROSUR program, the hotspots created in Italy and Greece, and the externalization of border control and migration management to other countries.[18] The focus on the interactions among actors, institutions, and technologies will cover monitoring technology to detect migration on land, at sea, and in the air, and the migration policies of the European Union and its member-states with countries on the southern and eastern sides of the Mediterranean.

The mobility of borders and their functioning as manifestations of policy and as vehicles for political thought can be seen in the various technological approaches to border control. While the aim of this discussion is not to provide an exhaustive historical overview of border politics in Europe, it will show how the relationships among politics, technology, materiality, and movement already inform actual developments in Europe's border politics. Building on this empirical foundation, chapter 3 will delve into a detailed conceptual discussion of the morphology of technopolitics.

Building Border Infrastructures

In 2015 and 2016, more than 2.5 million people applied for asylum in the European Union, while thousands of others lost their lives trying to get there, turning the Mediterranean into a veritable graveyard.[19] Also, in 2016, the ceiling of a museum in Schengen, Luxembourg, devoted to the history of Europe's passport-free zone collapsed during a storm, damaging installations and a number of exhibits, including passports, signed documents, photographs, and customs officials' uniforms. While this led to the museum closing for three months, Schengen's mayor was quick to note that the ceiling collapse was not a symbol. Nevertheless, "it is a sign that [they needed] to do some repairs."[20]

The use of the word "repairs" by Schengen's mayor suits the idea that the migrant crisis almost heralded the end of Schengen and of the circulation system that it initiated. But if we look at the Schengen Agreement from a different angle—namely, what happened inside the European Union, as well as what happened outside—its contemporary manifestation appears to be in line with its original aims. The creation of the Schengen Area—by now a visa-free zone spanning twenty-six countries—was (and remains) a way to order the relationship between Europe's internal and external borders. The relationship is often seen in terms of a dichotomy: an open European travel space on the inside and a closed border on the outside. But the Schengen Agreement and its many implementing policies introduced a mechanism to organize circulation that transcended the neat division between inside and outside. Its development was not only a matter of regulation, but also a thoroughly technopolitical matter.

It is tempting to take the fall of the Berlin Wall in 1989 as a turning point in how the borders in Europe were conceived. The opening of the Iron Curtain, the reunification of the two Germanys, and the dissolution of the Soviet Union can easily be seen as the beginning of a process often referred to as "rebordering": not the start of a new, globalized, borderless world, but a specific reorganization, relocation, and redefinition of borders of all sorts. Nevertheless, the process that would eventually lead to the distinction between Europe's internal and external borders started earlier, around 1984. These early policies were not in anticipation of migration from Africa and Asia that would reach Europe sooner or later, or were they made as a prelude to the migration that followed the war in the former Yugoslavia. The common border was instead part of a broader project—namely, the completion of the European internal market.[21]

On June 14, 1985, Belgium, the Netherlands, and Luxembourg joined Germany and France in signing a treaty to abolish their national borders. The signing of the Schengen Agreement was a moving event in many ways. The treaty was signed aboard the cruise ship *Princess Marie-Astrid*, which was moored near Schengen on the Moselle River. The Convention implementing the Schengen Agreement followed in June 1990. In 1995, the original signatories of the European Economic Community (EEC)—except for Italy—agreed to abolish internal border controls, while introducing a common visa system, greater law enforcement cooperation, and external border controls. The implementation agreement covered practical provisions, such as separate circuits for intra- and extra-Schengen flights at international airports. Following the signing of the Treaty of Amsterdam, the Schengen Acquis (the Agreement and Convention) was incorporated into the EU framework on May 1, 1999.[22]

Schengen's most remarkable invention was the distinction between internal and external borders. The Convention referred to internal borders as "the common land borders of the [Schengen states], their airports for internal flights and their ports for regular ferry connections exclusively from and to other ports within the territories of the [Schengen states] and not calling at any ports outside their territories." In contrast, external borders were deemed a Schengen state's "land and sea borders and their airports and sea ports, provided that they are not internal borders."[23] External borders were thus defined negatively.[24] The reason for doing so was to avoid the sensitive issue of who should be legally responsible for managing the border. Although this responsibility was never explicitly addressed, the text makes it clear who the responsible actors should be. When applied to internal borders, the term "common" refers to the territorial boundary shared by two countries. But when applied to external borders, it has a *collective* connotation. The result was that the abandoning of controls inevitably created a vacuum—a vacuum that had to be given content.[25]

To understand Europe's borders in terms of an infrastructure, the Schengen system must be viewed in relation to the Dublin system. To discourage asylum seekers from traveling to countries where they expect a more favorable reception, parallel to the implementation of the Schengen system, all EU member-states agreed in the 1990 Dublin Convention that, in principle, the member-state of first arrival would be responsible for examining applications for asylum. This Convention entered into force on September 1, 1997. It is important to note that the Schengen and Dublin systems were of an experimental nature. Both systems were established in separate treaties, outside the structure of the European Communities (EC). Schengen and Dublin, and particularly the technopolitical way in which they developed, can also be regarded as a form of infrastructural imagination. Schengen and Dublin not only came into being via a process of institutional competition, they also embodied a particular infrastructural innovation of actors, institutions, and technologies geared to developing novel border infrastructures. The Maastricht Treaty (1992), which entered into force in 1993, brought police, judicial, and migration cooperation under the umbrella of the European Union via the Third Pillar. Subsequently, the Treaty of Amsterdam (1997) integrated asylum and migration into the EC structures. From that moment on, the matter of Schengen and Dublin was controlled by EC regulations, supplemented by treaties with non-member-states.

The elasticity of Schengen and Dublin and the mechanisms that relate the internal borders to the external borders have been put to the test by various developments inside and outside the European Union. Internally, since 2004, the free movement of people gradually led to increased labor migration. From a legal point of view, this only concerned the functioning of the internal market, but in the official statistics of the member-states, as well as the perception of many citizens, these fellow European citizens were often regarded as foreigners. Externally, the end of repressive stability in North Africa and the Middle East had major consequences. Many people sought refuge in Europe as a result of the 2003 invasion of Iraq, the Arab Spring of 2010–2011, and especially the Syrian civil war that has been raging to some extent since 2011. Migration to Europe from Syria, North Africa, the Middle East, and parts of Asia increased in the period 2014–2016, particularly to Italy and Greece (discussion of the latter country is central to chapters 5 and 6). (The top countries of origin of asylum seekers were Syria, Iraq, Afghanistan, Nigeria, and Pakistan.)[26] This development was all the more intense now that many people had to take great risks when crossing through unsafe areas and had to expose themselves to exploitation by human traffickers. The number of dead and missing migrants in the Mediterranean that was recorded increased enormously.

The Schengen and Dublin joint area actually fell apart in countries on the external border that were under pressure from the arrival of entrants, especially Italy, Spain, Malta, and Greece, which were disrupted at the time by budgetary crises. Countries in Central and Eastern Europe did not favor immigration, and when immigrants did come, they offered them few prospects. In Europe, countries in Western and Northern Europe (Germany in particular) took in the lion's share of immigrants. The way in which the European Union attempted to stabilize these developments was driven by a harmonization of the procedures and criteria for admission and residence, building on the "acquis" of the Schengen and Dublin regulations. These procedures, however, were also underpinned by border infrastructure arrangements.[27] Two of the main developments concerning borders in the wake of the migrant crisis—the creation of so-called hotspots within the European Union and the externalization of border control outside it—were already anticipated in the initial proposals.

The European Union's border policies can be said to consist of all kinds of infrastructural compromises. The concept of "infrastructural compromises" bears witness to the entanglement of actors, institutions, and technologies, as it denotes that border infrastructures not only result from the negotiations among various political actors or the agreements between technical experts but they also arise out of all kinds of sociotechnical mediations.[28] In the context of border security infrastructures, compromises concern the transformation of conflicting requirements and opposing views into a workable composition by adding new elements, foregrounding certain aspects, and backgrounding others. The concept of an infrastructural compromise encapsulates both the materiality and the movability of borders, predominantly the exchanges and transportations that are required (in terms of knowledge, technology, and ideas) to constitute a border. As such, the perspective of border infrastructures not only allows the study of the particular technopolitical layout of borders, but also contributes to the understanding of technopolitical configurations as infrastructures and to the advancement of the study of sociotechnical mediation.

Compromises in the construction of border infrastructures also contain imaginations.[29] The particular aspect that the notions of compromises and imagination address is that there are mutual interactions between technology and politics that underpin the emergence of infrastructures. Borders and border politics are deeply implicated in materiality and movement. The notion of imagination emphasizes that political ideas are not developed separate from technology.[30] Imagination is required to invent new border control mechanisms to adapt to changing international circumstances and transformations in international human mobility. Border infrastructures themselves can be regarded as expressions of imagination that allow the creation of novel ideas and applications. As a toolkit for politics, border infrastructures are not just the instruments of politics. Instead, they encapsulate visions and ideas regarding the identity of Europe as a security actor. Not infrequently, these imaginations and compromises entail various contradictions and oppositions. These oppositions—such as between the inside and the outside of the European Union—have often created infrastructural tensions and fueled innovation in attempts to overcome them, as the following discussion will show.

Unpacking the European Union's Border Control Agencies

If it is true that border infrastructures as mechanisms to organize circulation are as much vehicles for political thought as instruments of political action, how are the entanglements between the European Union's institutions, agents, and technologies to be grasped? To comprehend the infrastructural characteristics of the entanglements between the European Union's institutions, agents, and technologies that were distinguished in the opening of this chapter, we must begin by conceiving of a parallel development of agencies and information systems.[31] This imperative can be followed by elaborating particularly on the first two characteristics of border infrastructures described earlier in this chapter— namely, that they connect large-scale networks with local situations and manifestations of borders and select by organizing forms of circulation.

The agencies that were installed include Europol, Eurojust, and, most important for the purposes of this chapter, Frontex. The European Police Office, which later became the European Union Agency for Law Enforcement Cooperation, and more colloquially known as Europol, commenced its full activities on July 1, 1999. From the outset, Europol was not directly concerned with migration, but it was involved with the prevention of human trafficking and the facilitation of illegal immigration as forms of organized crime. Eurojust, whose formal name is European Union Agency for Criminal Justice Cooperation, was established in 2002 and "supports in any way possible the competent authorities of the Member States to render their investigations and prosecutions more effective when dealing with cross-border crime."[32] The European Agency for the Management of Operational Cooperation at the External Borders of the Member States of the European Union (better known as Frontex) was agreed upon by European Council regulation on October 26, 2004, and established on May 1, 2005. It followed the December 2001 meeting in Laeken, at which the European Council called for a better management of the external border controls.

The name "Frontex" stems from the French *frontièr extérieur*. Arising out of seven ad-hoc centers on border control whose task was to oversee EU-wide pilot projects and common operations related to border management, the agency was created to fill the empty chair for the collective control over external borders.[33] Frontex rapidly emerged as the flagship of Europe's control over its external borders. Its primary task is to coordinate joint operations at the external land, sea, and air borders. Not only did Frontex build up a border control infrastructure, it also drew on existing systems of agencies such as the European Union Satellite Centre (EUSC), the European Defense Agency (EDA), the European Maritime Safety Agency (EMSA), the European Space Agency (ESA), and the European Centre for Disease Control (ECDC).[34] Although Frontex was involved in an institutional and infrastructural competition with Europol, the Agreement on Operational Cooperation between Europol and Frontex of March 28, 2008 expresses the intention to share information systems and to protect data.[35]

From its start, operations have been central to Frontex's mission. The first joint operation was Illegal Labourers in 2005. The first sea operation began in 2006 and was called Hera. Hera was concerned with irregular immigration from West Africa to Spain's Canary Islands in the Atlantic Ocean. It soon became clear that the Mediterranean would become the main area for joint operations. The shift of focus to the Mediterranean not only implied a particular operational angle, but also introduced an additional motif— namely, the requirement of *interoperability*, a concept that will be discussed later in this chapter. When Frontex began its mission, there was little international cooperation in the sphere of external border control. One of the agency's headline initiatives was the establishment in 2007 of Rapid Border Intervention Teams (RABITs), the first of which was deployed in 2010 along the Evros River on the Greek-Turkish border.

The goals of EU policies with regard to international migration consist of a combination of mobility, security, fundamental rights, and humanitarian considerations. In all the key documents, the European Parliament, the European Commission, and the Council of Ministers underline that the European Union ought to commit to preserving life at sea, combating human smuggling and trafficking, and respecting refugee rights.[36] When Frontex was originated, it lacked a concrete human rights framework. Border security considerations such as assisting national border guards, coordinating joint maritime missions, and coordinating EUROSUR were defined as its main task. On the other hand, as an EU agency, Frontex is bound by the EU Charter, indirectly by the European Convention on Human Rights and the jurisprudence of the European Court of Human Rights (ECHR), and also by international human rights standards and protection obligations toward asylum seekers. An important underlying idea is the principle of "nonrefoulment," which forbids a country receiving asylum seekers from returning them to a country where they are in danger.

Still, a gap needed to be filled in order to do justice to the tensions arising between security considerations and fundamental rights. In 2014, the European Union formulated the Frontex Sea Borders Regulation (regulation 656/2014). This regulation replaced the previous Council Decision 20120/252/EU; it aimed to establish "clear rules of engagement for joint operations at sea, with due regard to ensuring protection for those in need who travel in mixed flows, in accordance with international law as well as increased operational cooperation between countries of origin and transit."

Although the regulation sidesteps some political controversies with regard to refugee rights, immigration deterrence, and saving lives at sea, it is a stronger instrument than the previous Council Decision.[37] Regulation 656/2014 of the European Parliament and the Council of May 15, 2014, establishing rules for the surveillance of the external sea borders in the context of operational cooperation coordinated by Frontex, introduced changes to the mandate of the agency. Together with the earlier Regulation 1052/2013 of October 22, 2013, which established EUROSUR, it was fully integrated and referred to in Regulation 2016/1624 of September 14, 2016, which installed a "European Border and Coast Guard Agency" to replace the "European Agency for the Management of Operational Cooperation at the External Borders of the Member States of the European Union."

The development of these regulations is significant here, as it shows how the European Union and Frontex aimed to combine the different but inevitably related tasks of protecting borders and people. The legal nuances will not be discussed at length. Instead, the focus will be on the attempts to combine border control, security considerations, and human rights in terms of the construction of "infrastructural compromises" referred to in chapter 1. An infrastructural compromise does not result from an agreement between political actors; rather, it emerges from infrastructural tensions among actors, institutions, and technologies. Chapters 5 and 6 will detail the constituent components of such a compromise and the ways that it worked out—or failed to work out—in practice. Moreover, these chapters will demonstrate how the quest for infrastructural compromises fuels the multiplication of the border. Infrastructural compromises intensify the mediating nature of borders and push them toward novel hybrid connections, such as between surveillance on land and at sea, or between care and control.

The background of these compromises consists not only of legal struggles, political negotiations, and policy considerations, but also of conflicts among various actors, institutions, and technologies in particular situations and specific circumstances. These conflicts flared up by a series of dramatic events at sea and the increasing number of migrant deaths in the Mediterranean. Chapter 7 describes in detail how migrant deaths and dramatic events at sea become public issues and discusses questions of responsibility and accountability that these events evoke. The chapter will do so by discussing the notion of the infrastructural state and the rise of the observing, investigating, reconstructing, and participating public eye. One of the events includes the so-called left-to-die boat, a widely documented case.[38] The term "left-to-die boat" refers to an event in March 2011 when sixty-three migrants lost their lives in the central Mediterranean while attempting to migrate from the coast of Libya to the small Italian island of Lampedusa.[39] What made the incident particularly unfortunate is that these deaths occurred when the European Union's maritime frontier was under high surveillance. National border police forces from both sides of the Mediterranean were reinforced by over forty military ships and many patrol aircrafts deployed by Western states off the Libyan coast in support of international military intervention led by the North Atlantic Treaty Organization (NATO). This high density of surveillance at sea "placed these deaths squarely in the most highly surveyed waters in the entire world, and there were strong indications that military forces were failing in their obligation to rescue migrants in distress, despite possessing the requisite means of surveillance to witness their plight."[40]

The causal relations between such dramatic events, the transformation of public opinion, political responses, and institutional and operational change are hard to determine (if they exist at all). In some cases, dramatic events are a consequence of measures taken—or not taken—by the European Union, its member-states, and/or other European countries. In other cases, these dramatic events resulted in new policies and regulations to strengthen the protection of migrants. Over the past decade, hundreds of dramatic events resulting in thousands of deaths have taken place. Next, a brief and far from complete overview of some of these cases will be presented. The cases are selected because they had a specific impact on the development of policies and regulations, became events that affected public opinion at large, or both.

The first case took place in 2009, when the Italian government supported pushback operations to prevent migrants from reaching the Italian islands. It thereby withheld a proper refugee status determination procedure from the migrants. Twenty-four migrants who were part of a larger group intercepted in May 2009 in an Italian pushback operation and returned to Tripoli brought their case to the ECHR. In 2012, the Court formulated the *Hirsi* judgment and considered the pushbacks a clear violation of the rights of the migrants. These issues were not fully solved with Frontex's 2014 regulation, which halfheartedly incorporated the standards of the judgment.[41]

The aforementioned left-to-die case of 2011 gave rise to an investigation of the Parliamentary Assembly of the Council of Europe.[42] The Lampedusa shipwreck tragedy on October 3, 2013, which caused the death of 368 migrants, was a turning point for both Italian and EU policies. The Italian government launched Operation Mare Nostrum, while the European Commissioner for Home Affairs called for the European Union to increase its Mediterranean-wide search-and-rescue patrols to intercept migrant boats through Frontex. A significant pushback disaster in the Greek Aegean Sea is the Farmakonisi case of January 20, 2014, which will also be discussed in chapter 7, in which eleven refugees—eight of them children—lost their lives when their boat capsized as it was being towed through the water. The event inspired director Anestis Azas to write and direct a performance called *Case Farmakonisi or the Justice of the Water* in 2015. On April 18, 2015, at least 800 people died as their boat capsized between Libya and Lampedusa when it collided with a Portuguese freighter ship that had been called to its aid. Only twenty-eight people survived the accident. The disaster occurred after the ending of the Mare Nostrum project, in which distressed boats carrying migrants were rescued at the expense of Italy and the European Union. (The shipwreck, one of the deadliest in the Mediterranean in living memory, was displayed in an exhibition by the Swiss-Icelandic artist Christoph Büchel at the 58th Biennale di Venezia in 2019.)[43] The disaster accounted for nearly a fifth of the estimated 3,665 migrant deaths in the sea that year. In response, at a joint meeting of foreign and interior ministers chaired by High Representative/Vice President Federica Mogherini and held in Luxembourg on April 20, 2015, Migration, Home Affairs, and Citizenship Commissioner Dimitris Avramopoulos presented a ten-point plan of the immediate actions to be taken in response to this crisis in the Mediterranean.[44]

On April 28, 2015, the European Commission adopted the European Agenda on Security, followed by the European Agenda on Migration on May 13, 2015. The near-joint publication of the two agendas states that external border management is increasingly understood as a pact to organize the inflow of migrants and internal security.[45] In addition, the European Council agreed to reinforce the scope of the EU civilian mission in Niger to support the authorities in preventing irregular migration and combating associated crimes, as well as to establish an EU military operation, EUNAV-FOR Med, to break the business model of smugglers and traffickers of people in the Mediterranean. In June 2015, EU leaders agreed on a series of measures covering the areas of relocation and resettlement, return and readmission, and cooperation with third countries (meaning countries that are not a member of the European Union, as well as countries or territories whose citizens do not enjoy the European Union right to free movement).

A second package of proposals in September 2015 included an emergency relocation proposal for 120,000 people in clear need of international protection from frontline countries; a controversial relocation mechanism for all member-states; a highly disputed common European list of safe countries of origin; a supposedly more effective return policy; a guide on public procurement rules for refugee support measures; measures to address the external dimension of the migrant crisis; and a trust fund for Africa. As mentioned later in this chapter, these measures were followed by the Valletta summit and the EU-Turkey Statement.

Not only have the attempts to combine border control, security, and human rights approaches with regard to migration resulted in all kinds of infrastructural compromises (as chapters 5 and 6 will explain in greater detail), but Frontex itself increasingly has become a paragon of infrastructural imagination. Proclaiming operations as Frontex operations, the fluttering EU flags on the participating vessels, and the Frontex armbands worn by the officers create an impression to migrants that it is Frontex, not the individual member-states, that they are dealing with. The growth of the agency and the broadening of its tasks make it a symbol of European cooperation.

In 2016, Frontex was rechristened the European Border and Coast Guard Agency (EBCG). But in addition, the role of the EBCG is entirely new; it was never part of the original Frontex mandate.[46] EBCG can intervene directly in the affairs of a member-state that proves ineffective in controlling its own borders. An important shift brought about by the EBCG regulation is the introduction of a centralized mechanism that deals with situations where control of the external border is rendered ineffective. This would be the result of inadequate measures by the member-state to prevent irregular access, which could result in jeopardizing the functioning of the Schengen Area as a whole.[47] It can also collaborate with the Commission in a hotspot at a member-state's request.

The position of Frontex still raises many questions. The increased capacities, tasks, and competences of Frontex have been accompanied by new accountability mechanisms, but the exact meaning of these mechanisms in the organization's operations is far from clear. On November 13, 2019, the EBCG was given a new legal basis.[48] In addition to the broadening of its mandate, the new regulation offers Frontex its own standing crops, equipment, and a greater role in the governance of border surveillance data systems. The extension of the EBCG's mandate, staff, and equipment suggests that Frontex will play an even larger role in Europe's border management—and it will. But Frontex is not the engine of Europe's border machine. The development of Frontex as an agency mirrors the changing nature of Europe's borders and the continent's border control as well.

Digital Borders: The European Union's Databases and Information Systems

Of the characteristics of border infrastructures that were discussed in the beginning of this chapter, the analysis so far has mainly focused on the way that border infrastructures connect large-scale networks with local situations and manifestations of borders, and how border infrastructures choose between migrants by organizing forms of circulation. Borders, however, also display a specific interplay between visibility and invisibility. This interplay is expressed in particular by the relation between the aforementioned border infrastructures and Europe's digital borders.

Alongside the aforementioned policies and agencies, the European Union has developed three information systems: SIS, Eurodac, and VIS. These systems are operated by eu-LISA (the much more manageable name for the European Union Agency for the Operational Management of Large-Scale IT Systems in the Area of Freedom, Security and Justice). The databases and information systems are the digital technological dimension of Schengen and the Dublin Regulation. However, it would be misleading to consider these databases as only digital or virtual information systems. As the following chapters will show, the systems are directly related to the border practices on land, at sea, and in the air, and also affect the border infrastructures of harbors, airports, and hotspots.

SIS is a large-scale information system that supports external border control and law enforcement cooperation in the Schengen states. According to the Commission, it is the "most widely used and largest information sharing system for security and border management in Europe."[49] The Schengen states were early in developing an electronic mail infrastructure. SIS monitors all kinds of cross-border movements. As the "database-flagship" of the Schengen Agreement, it stores information on persons and objects and enables national authorities such as the police and border guards to enter and consult alerts about them.[50] SIS consists of a central database, called C-SIS, and national SISbases, in all of the Schengen states. C-SIS is located in a bunker in Strasbourg, and its aim is to maintain "public order and security, including State security, and to apply the provisions of this convention relating to the movement of persons, in the territories of the contracting parties, using information transmitted by the system."[51]

In April 2013, the second generation of the Schengen System (SIS II) went live. SIS II consists of a central system (Central SIS II), a national system (N.SIS II) in each member-state, and a communication infrastructure that links Central SIS II to the various national systems. SIS II enables authorities such as police and border guards to enter and consult alerts on certain categories of wanted or missing persons and objects. The reasons for issuing an alert include to refuse entry to a person who does not have the right to enter or stay in the Schengen territory, to find and detain a person for whom a European arrest warrant has been issued, to find a missing person, and to find stolen or lost property, such as a car or a passport. An SIS II alert contains not only information about a particular person or object, but also clear instructions on what to do when the person or object has been found.

As the European Data Protection Supervisor (EDPS) describes, the competent authorities of the member-states enter, update, or delete data in the SIS II via their national systems. Before a competent authority issues an alert, it has to determine whether the case is relevant enough to warrant entry. The competent authorities are also responsible for ensuring that the data is accurate, up to date, and lawfully entered into SIS II. When the alert is issued in SIS II, only the relevant member-state is authorized to modify, correct, update, or delete the data. According to the Commission, by the end of 2017, SIS contained approximately 76.5 million records, and it was accessed 5.2 billion times and saw 243,818 hits.[52]

The Eurodac system collects the fingerprints of asylum seekers in support of the Dublin Regulation. Acting as the EU asylum fingerprint database, it charts asylum migration across Europe and detects multiple asylum claims. Eurodac was introduced to prevent so-called asylum-shopping. Through the registration of multiple claims, it also gives some indication of secondary movements. The biometric information database became operational in 2003. The system consists of a central unit, a computerized central database used to compare the fingerprint data of asylum applicants, and the means of data transmission between the member-states and the central database. The EDPS is responsible for supervising the system in cooperation with the competent national data protection authorities. According to this entity, "when a participating country sends a set of prints to Eurodac, it knows immediately if they match up with others already on the database. If so, it can choose to send the individual back to the country where he or she first arrived or applied for asylum; the authorities there are responsible for making a decision about the candidate's right to stay. If not, the country that submitted the prints handles the case."[53]

VIS became operational in October 2011. It connects EU member-states' immigration authorities and consular posts around the world. Its central database details the personal and biometric information of all visa applicants to the European Union and the dates on which the visas were applied for, granted, refused, cancelled, withdrawn, or extended. This means that VIS generates aggregated data on shifts and trends in visa-based mobility from specific countries and regions.[54] VIS stores and processes three categories of data: (1) alphanumeric data on Schengen visa applicants and holders (e.g., surnames, names, and places and dates of birth), as well as on each specific application (e.g., status of the application, the authority processing it, and the type of visa requested); (2) authorities capture, digitize, and store in the system the ten fingerprints of each applicant; and (3) photographs (i.e., facial images) of those requesting Schengen visas. In addition, VIS contains scanned documents submitted by individuals in support of their visa applications.[55]

These technologies have raised many ethical and legal questions. Processes of identification, authentication, and registration increasingly rely on information derived from human bodies, such as fingerprints, and thus they may invade people's privacy and affect how they view their bodies. Such interventions place high demands on the integrity of companies and professionals and prevent overly restrictive interpretations of the tests lead to questionable rejections and passes. Gathering information extracted from human bodies also supports dragnet policies in which vast amounts of data are collected for future purposes. Eurodac, for instance, was originally intended to prevent multiple asylum applications and unauthorized entry. Later, under renewed Eurodac regulations, access was no longer restricted to immigration authorities, but widened to include police, public prosecutors, and Europol.[56]

Besides specific ethical, technical, and financial concerns, a particular criticism is about the capacity of these systems to become "ever closer," to interconnect and expand. For these reasons, these systems have been labeled as "greedy" and as a "machine."[57] This characterization is particularly applicable, given EU proposals regarding the transformation of these systems as a response to the so-called migrant crisis and the ongoing intermingling of migration and security policies. In April 2019, the European Union adopted new legislation with regard to its digital borders (namely, legislation to establish a framework for interoperability between several EU information systems). This legislation affects the areas of security, border, and migration management, visa processing, and asylum because it concerns the Schengen acquis regarding borders and visas, as well as the Schengen acquis on police cooperation. The legislation concerns VIS, Eurodac, and SIS, as well as three databases that do not exist yet—the Entry/Exit System (EES), the European Travel Information and Authorization System (ETIAS), and the European Criminal Records Information System for Third-Country

Nationals (ECRIS-TCN). It also concerns Interpol's Stolen and Lost Travel Documents (SLTD) database and Europol data.

The EES aims to "contribute to the modernisation of the external border management by improving the quality and efficiency of the external border controls of the Schengen Area."[58]

ETIAS ought to become "an automated system that would gather information on visa-exempt travelers prior to their arrival, in order to determine any irregular migration, security or public health risks associated with them."[59] ECRIS-TCN "aims to improve the exchange of criminal records information regarding convicted non-EU-citizens and stateless persons through the existing European Criminal Records Information System."[60] The interoperability among these systems consists of four components: (1) a European search portal (ESP), (2) a shared biometric matching service (shared BMS), (3) a multiple identity detector (MID), and (4), the eye catcher of this interoperability operation, the Common Identity Repository (CIR). Final approval of the Commission's proposal to create a European Criminal Records Information System for Third Country Nationals was given by the Council on April 9, 2019.[61]

The legislation has been widely criticized. This criticism will sound familiar to scholars interested in issues of border surveillance. The nonprofit organization Statewatch published a report in 2018 titled *Interoperability Morphs into the Creation of a Big Brother Centralized EU State Database Including All Existing and Future Justice and Home Affairs Databases.*[62] A German Member of the European Parliament described the development as creating "a monster database." The European data protection supervisor has warned of a potential "panopticon in which all our behavior is considered useful for investigative purposes and must be made accessible because fighting crime is given priority."[63]

Without wanting to deny these qualifications, there is a different issue at stake. The example of the aforementioned legislation is interesting in itself, as it raises all kinds of questions and has far-reaching consequences, but it also illustrates more general characteristics of digital borders in particular and border infrastructures in general. A particular characteristic of border infrastructures concerns the morphology of borders. "Morphology" here means the study of the development and dissemination of political ideas via their manifestation in all kinds of technologies, as well as how, via those expressions, the ideas themselves are affected and transform, grow, and change. In this morphological conception, border control technologies (such as walls and fences) or border surveillance technologies (such as large-scale databases) are not just instruments or tools that are freely available and acted upon at will. Instead, the following discussion will explore the idea that border politics develops through all kinds of technologies and expresses itself in the shape of borders.

An emphasis on morphology allows a consideration of the shape and the form of borders, technologies, and politics. Two aspects require special attention: the movements that shape borders and the materiality that makes them. In that sense, the present view shares some similarities with the notion of "kinopolitics," according to which "a border is not simply an empirical technology to be resisted or not; it is also a regime or set of relations that organize empirical border technologies."[64] The argument that will be set out here can also be regarded as relational. However, an even more intimate relationship between knowledge, ideas, politics, and technology seems to be at stake. For this reason, border can be conceived as a kind of containers or vehicles of politics. This does not mean vehicles like remote-controlled cars that can be acted upon at a distance, but rather vehicles as moving entities that contain and transport ideas and imaginations of Europe, of belonging, of identities, of inclusion, and of exclusion. In a comparable way, the forthcoming database that is part of the legislation, the CIR, can be regarded as a vehicle that not only stores information, but also gathers it, distracts it from humans, and mobilizes it for various purposes.

EUROSUR has been developed following a different institutional and infrastructural logic than the aforementioned databases. The course of the European Commission with regard to the mandate, funding, and staffing of Frontex was informed by two studies—the MEDSEA and BORTEC studies. The MEDSEA attended to the possibilities for enhancing operational cooperation in the patrolling of the European Union's southern maritime borders and the Mediterranean. The study provided for the launch of Mediterranean coastal patrol networks and information-sharing mechanisms between the member-states and FRONTEX. The BORTEC study, on the other hand, was concerned with the setup of a European border surveillance system focusing on the EU southern maritime borders, including the Mediterranean.[65] The BORTEC report is a telling illustration of the way that persons are increasingly regarded as subjects that are part of large-scale populations and need to be objectified as "targets."[66] The plans presented in the BORTEC study eventually led to EUROSUR, which became operational in December 2013.[67]

Instead of creating an all-seeing eye or a seamless web, the coupling of different technologies leads to a combination of systems. Monitoring mobility requires protocols and personnel to gather, interpret, compare, and apply information. This monitoring is based on a distinction of different areas, such as the coastal waters of EU member-states, the open sea, and coastal waters of third countries. Each area requires specific modes of detection, such as systems to identify vessels by monitoring and tracking, radio, coastal radar, infrared cameras, satellites, and unmanned aerial vehicles (UAVs) such as drones. The key word is "interoperability," and the aim of such monitoring is to create situational awareness. A military term by origin, situational awareness aims to visualize critical situations such as emergencies and irregular border crossings to assess whether intervention is required. In addition to boats, cameras, and radar, since 2014, EUROSUR has been using satellite imagery obtained through the European Satellite Centre. For example, when a Hellenic coast guard patrol spots an unregistered ship in Greek waters, it contacts the national coordination center in Piraeus, which directs it to the agency's headquarters in Warsaw to compare the crew's observations with satellite images. Armed with this information, the coast guards can then decide what to do.

The notion of interoperability, as this discussion shows, applies to the interconnection of the European Union's various databases and information systems concerning migration, borders, and security, as well as to the cooperation that is required for specific border operations, such as the ones conducted by Frontex and the EU member-states. The EDPS says, with regard to the new legislation that was adopted by the European Union in April 2019, that "interoperability is not primarily a technical choice; it is first and foremost a political choice to be made."[68] A political choice indeed, but of what kind of politics? The issue of pursuing the policies of interoperability is not just that a big, greedy data monster or an all-seeing apparatus is created. Most striking is the appearance of all kinds of novel mediating moments at which new connections are being established among actors, institutions, and technologies.

The Technological External Dimension

The external dimension of Europe's border control illustrates the fourth characteristic of border infrastructures—their movability—par excellence. Externally, the European Union—particularly in the form of its neighborhood policies—has become increasingly involved in cooperation, development, security, and migration management, implying the possible movement of the external frontier as well. For instance, the Barcelona Process, set up in 1995 to coordinate relations between European countries and their North African and Middle Eastern neighbors, resulted a decade later in the Euro-Mediterranean Summit, held in Barcelona. The Barcelona Declaration merely adapted the process to new challenges, such as with regard to migration. The externalization of border control and mobility management is an important and growing part of EU foreign policy, particularly concerning its neighbors to the south and east. Externalization of border control establishes that borders can be movable entities themselves. Initiatives to support democracy and development have increasingly become part of an agenda that is engaged primarily with security and considers migration issues and border control policies accordingly. The externalization of border control also opens a new technological dimension.

Spain's policies during the 2000s, which to a certain extent foreshadowed the European Union's Global Approach to Migration, offer a telling example. After closing the transit routes via the Strait of Gibraltar and the Ceuta and Melilla enclaves in 2005, migration moved to the Canary Islands. Spain subsequently sought to strengthen control in West Africa. Spain's policies were written down in *Plan Africa,* published by the government in 2006. At the core of Spain's approach was the connection between development policies and migration governance. The plan was enacted through the deployment of various technological operations. First, Spain intensified its patrol of the waters surrounding the Canary Islands. Spanish marine and Guardia Civil (Civil Guard) vessels, assisted by planes, surveilled the waters to detect migrants' boats. Thereafter, in 2006, Spain established new treaties with the government of Mauretania. Together with Spanish police and Guardia Civil officers, they brought along surveillance equipment, including a helicopter with a night observation device, a surveillance plane that was handed over to Mauritian forces, and joint patrols of Spanish and Mauritian security forces in the harbors and coastal areas. Mauritian forces, including coast guard, were trained by Frontex standards and equipped with vessels, zodiacs (fast rubber boats), quads (small four-wheel vehicles), and surveillance technologies.[69]

This remote control consists of moving border controls farther and farther south, east, and southeast, away from the boundaries between neighboring European countries. Since the Tampere Council in 1999 placed the spotlight on the external dimension of the European Union's migration and asylum policies, strategies have emphasized mobility partnerships, partnership with countries of origin, and stronger external action. Since 2005, many of these external aspects have been managed by the European Union's Global Approach to Migration and Mobility (GAMM), while the European Agenda on Migration, formulated in 2015, has been particularly important in addressing the rise of migration since 2014. The agenda is based on four pillars: (1) reducing the incentives for irregular migration, (2) border management, (3) a common asylum policy, and (4) a new policy on legal migration. The agenda in this sense aims to arrive at a compromise, a combination of migration and security polices and humanitarian initiatives. At the international summit in Valletta in November 2015, European and African leaders agreed to intensify remote control. Measures included new legislative and institutional frameworks to ensure the control of land, sea, and air borders and the provision of equipment, anti-trafficking training, and intelligence services.[70]

The European Union carries out its border work far beyond the external borders of the current union.[71] The New Deal for Africa, as the European Union's Partnership Framework has been called, aims to invest tens of billions of euros over the coming years into a range of financial instruments and funds, most notably the EU Trust Fund for Africa, designed to create jobs and strengthen the communities, policies, and border controls in Jordan, Lebanon, Niger, Tunisia, and elsewhere. The initiatives range from coordination among regional land and sea border control authorities, such as the Seahorse network across West Africa, to investment and development programs aimed at limiting migration and combating international crime and terrorism.

The focus on managing migration and externalizing border control in the European Union's common foreign policy has led to its creeping securitization, which has only been exacerbated by the EU response to the migrant crisis. On November 7, 2015, on the eve of the migration summit in Malta, the Dutch minister for foreign affairs stated that "the migration agenda demands serious cooperation with Africa. Border controls, terrorism, smugglers' networks: there is scope for compromise on all these issues." In his view, the upcoming summit offered the prospect of a "New Deal" between Africa and the European Union.[72]

The deals reached by the European Union with countries to the south and east of the Mediterranean combine migration management with issues of diplomacy, trade, development, and security. The EU-Turkey Statement holds that from the day of the agreement, all new "irregular migrants" crossing from Turkey to the Greek islands will be returned to Turkey. This should be done after an assessment of each individual's asylum claims, in line with EU and international law. For every Syrian returned to Turkey, another Syrian would be resettled directly from Turkey to the European Union.

The EU-Turkey Statement has been roundly criticized. Unclear whether it can actually be considered a treaty, its legitimacy remains in doubt. The current situation in Turkey makes it questionable whether the principle of nonrefoulement in international law can be respected. The effectiveness of this statement has been questioned as well, with numerous scholars pointing out that arrivals were already declining, partly due to Hungary's closing of the Balkan Route. The execution of the agreement also has left much to be desired: returns to Turkey are processed slowly, as are requests for asylum. Many migrants thus remain trapped on the islands in centers and camps. Numerous critics also argue that the statement has made the European Union vulnerable to political blackmail and has strengthened relationships with a country on an illiberal slide. The statement disregards the procedure laid down in the EU treaties. For that reason, it prevented the European Parliament and the European Court of Justice to take their constitutional roles. At the individual level, the statement was hard to challenge by those whom it affected.[73] Moreover, it blocked their access to the legal system.[74]

So-called deals like the EU-Turkey Statement do not consist of only international treaties, political agreements, policy mechanisms, and funding. As the previous examples show, border externalization contains a technological dimension as well. This dimension becomes particularly clear in the EU policies with regard to North Africa and sub-Saharan Africa. The volume of the technological container of border externalization has increased in particular by the intermingling of two agendas: the EU migration management and security agenda on the one hand, and making a compromise between border control and protecting lives on the other. This intermingling was addressed in particular in the 2015 European Agenda on Migration and the twin document, the 2015 European Agenda on Security. According to the European Union, the two agendas ought to be read together. Many of the tools, instruments, and devices that are developed for the purpose of migration policies, such as biometric information on identity and travel documents and risk assessments coincide with the EU security strategy.[75] These technopolitical tools are powerful performative devices facilitating the securitization of migration.[76]

The complications among Italy, Libya, and the European Union and its other member-states reveal various examples of the technological external dimension and the inseparable connection of this dimension with borders and human lives. First, to prevent uncontrolled migration and save lives in the Mediterranean, Italy and the European Union provided Libyan coast guards and migration control officers with training and instructions. Second, the technological dimension consists of the provision of vessels and the organization of joint patrols. Third, it contains the use of radar systems, satellites, and drones, as well as cameras and infrared sensors installed on ships, high-resolution binoculars, and night vision equipment, perhaps to be extended with software to track and identify ships.

The technological dimension consists not only of the construction of all kinds of tools, instruments and apparatuses, but also the destruction of things. Destruction can be understood in the literal meaning, such as the destruction of boats by the EU Naval Force Mediterranean that were considered to be used for the transport of migrants in the period 2015–2017. But destruction also concerns rights and values. In 2017, the Italian government aimed to restrict the activities of organizations trying to rescue migrants at sea. The Italian government has created a monopoly to conduct search-and-rescue operations—or to refrain from them.[77] In 2018, the Italian government prevented the docking of ships at ports that transported rescued migrants.[78] In addition, the Italian Maritime Rescue Coordination Centre warned the Libyan coast guard in order to allow them to rescue people and return them to Libya.[79] This situation created pullbacks (i.e., "remote control pushbacks").[80] In 2019, the Italian government passed a security decree that criminalizes search and rescue and humanitarian aid, allowing it to fine NGOs and migrant rescuers.

Although the previous examples mainly refer to the Italian government, humanitarian aid to migrants is under pressure throughout Europe. The report of the United Kingdom's Institute for Race Relations (IRR), *Humanitarianism: The Unacceptable Face of Solidarity*, offers many examples of legal and political suppression of support for migrants.[81] The suppression ranges from the banning of volunteers distributing food to proferring legal charges against people offering shelter to migrants.[82] All these examples show that there is a specific material aspect involved in the European policies of security and border externalization that lead to various forms of construction, destruction, criminalization, and containment.

Europe's involvement is not restricted to the countries and coast of North Africa; they reach even deeper, into the African continent. The combination of migration, development, and stabilization policies affects sub-Saharan countries, as well as countries in the Middle East. The New Migration Partnership Framework of the European Commission, passed on June 7, 2016, proposed Ethiopia, Jordan, Lebanon, Mali, Niger, Nigeria, and Senegal as its priority partners, but the collaboration also concerns other African countries such as Libya and Sudan. In practice, the European Union's involvement leads to cooperation not only with governments and elected representatives, but also with various political and military groups and local militias and clans or state security organizations that are not known for their protection of human rights.

The European Union collaborates closely with the International Organization for Migration (IOM). In 2009, the IOM developed the Migration Information and Data Analysis System (MIDAS), "a high-quality, affordable system that can collect, process and record information for the purpose of identification of travelers, data collection and analysis."[83] MIDAS is installed in over 100 land, air, and sea border-crossing points in twenty nation-states in Africa and the Americas. IOM also encourages the use of biometrics in migration management. Between 2012 and 2016, IOM offices in eighty countries had implemented 125 projects with significant biometric components.[84] These technologies do not affect just migrants who are on their way to Europe. By far, most of the migration movements are located within and between African countries themselves. As a result, these technologies also affect people conducting seasonal labor and communities that are not bound to state borders. International interactions and transactions set the agenda and intensify the use of technology. However, not all technological border infrastructures developed in Africa are supported by European Union or the IOM. Another driver of the emergence of border infrastructures consists of the circulation of knowledge about border control technologies by transnational security professionals (e.g., with regard to biometric security practices in Senegal).[85]

The externalization of border control not only displaces the border, it also creates novel interactions between the inside and the outside of Europe. By moving the border outward, border infrastructures increasingly become mechanisms that not only organize circulation, but support security and sta-bility as well. As a result, the border multiplies: it involves an increasing num-ber of actors to develop it, as well as an increasing number of people to which it is to be applied. The external technological dimension adds another layer of complexity to the inside-outside relationship. On the one hand, it enlarges what is considered the inside of Europe as border infrastructures extend. On the other hand, the inside-outside division only becomes harder to maintain as the multiplicity of border practices shapes ever-more-entangled relation-ships. In that sense, the externalization of border control via technologies intensifies the emergence of mediating moments—and, when we consider the role of patrol boats, helicopters, radars, checkpoints, harbors, and bio-metrics, we may add material moments, as they were described earlier in this chapter. These moments multiply the construction of compromises, such as between development and stabilization, and security and humani-tarian concerns.

Technological Borders: A Laboratory of Europe

At first glance, the development of Europe's borders reveals the logics underlying European cooperation and integration—namely ongoing inte-gration leading to spillover effects that are addressed by creating suprana-tional institutions. This looks like the teleological logic of the European Union in optima forma. Indeed, the Schengen Agreement of 1985 and the 1990 Convention implementing the Schengen Agreement were the result of increasing cooperation in the form of the internal market and freedom of movement, which prompted the EU member-states to harmonize their border policies. But a closer look shows things to be more nuanced, and in many cases, it displays the outcome of subtle negotiations within chang-ing historical and technological contexts, rather than the result of a well-planned integration process.[86]

The previous sections discussed the characteristics of borders as infra-structures as introduced in the opening of this chapter. First, borders con-nect large-scale networks with local situations and manifestations of borders. Second, borders select among migrants in particular ways—not just by including some and excluding others, but by organizing forms of circulation. Third, borders display a particular interplay between visibility and invisibil-ity. Fourth, borders can be movable entities themselves. All together, these features represent the development of Europe's borders as an infrastructural laboratory. The term "laboratory" is referred to here in two ways. In a historiographical meaning, the notion of the laboratory was used by Schengen officials and EU officials as a metaphor to emphasize the experimental and innovative nature of the initiative outside the formal institutions.[87] Besides this metaphorical meaning, laboratories can be understood in a conceptual sense, as is frequently done in political science and technology studies. In this conceptual meaning, a comparison is drawn between the laboratory and the state—or, in the case of the European Union, a union of states. Not unlike laboratories, states consist of various actors capable of mobilizing each other and forming associations to execute specific tasks.[88] States or a union of states concerned with its internal and external borders, must experiment and test their own programs of action. In that sense, like a laboratory, the European Union is a setting where issues and experiments circulate from the micro to the macro level and back.[89] As such, the analogy envisages the EU actors and agencies as mediators that can enroll a network of instruments.[90]

Meanwhile, the notion of the laboratory is not an innocent metaphor. The metaphor is a very demanding one, in that in the context of borders, technologies, and politics, it implies that experiments are conducted on humans—their lives and rights. The reason that the laboratory metaphor can still be considered appropriate is that, besides the fact that it has been used regularly by EU officials to typify the role of "Schengen," the notion of a laboratory is much more specific than the metaphorical meaning of a test lab. A laboratory is not just a test lab, but an entity that denotes a specific way of organizing innovation and entails a specific constructivist view to describe this innovation process. In the case of the European Union and Schengen, the innovation consisted of the novel relationship between internal and external borders and the coordination and organization that was required to control them simultaneously. The Schengen Agreement and the Schengen Convention were adopted outside the framework of the European Communities (later the European Union) in 1985 and 1990. The Schengen Protocol was attached to the Treaty of Amsterdam in May 1999.

In the sense of being a laboratory, Schengen was part of a competition, an infrastructural composition to use the terminology of this study. Commenting on the early stages of the initiative in a 1990 report on "the removal of controls on persons at the internal frontiers of the Community," the Commission of the European Communities defined Schengen as "an exercise" that would function as a "testing ground" or "test bed" for what would happen in the European Union. This metaphor was also used by the former general director of the Justice and Home Affairs Directorate, when he noted that "the proponents of Schengen are not working in vain; they are demonstrating a possible and feasible way, creating a laboratory for Europe, and ultimately offering a decisive push to the European construction."[91]

Europe's infrastructural practices that help constitute the border share many similarities with laboratories in a conceptual way. The collection of rules, regulations, and treaties often referred to as "Schengen" concerns much more than the creation of an open European space. Schengen created this space, but it did so by redefining internal and external borders in a very specific way. Moreover, Schengen is not only concerned with Europe's internal space.[92] European border practices did not disappear with the Schengen Agreement. Schengen's main invention was not just the distinction between internal and external borders, but also the creation of a kind of political accordion that allows the squeezing in and drawing out of border policies. The idea behind the open European space and the intensified control of external borders was neither to close the European Union off nor to not allow anybody in any longer, but rather to create a filtering system, a sieve for selection. Instead of creating an all-seeing eye or a seamless web, the coupling of different technologies has led to a combination of systems, with many gaps between them.

Europe Inside Out, and Outside In

The division between internal and external borders has transformed the nature of borders and led to a complex relationship between what counts as inside and outside the European Union. Chapter 3 will proceed with a conceptual, morphological exploration of the relation between inside and outside, between concepts, ideologies, and ideas and between materialities and technologies of all sorts. To lay the groundwork for exploration, the following lessons can be drawn from this chapter.

The first aspect of Europe's borders is that they cannot be considered as plain boundaries between a territory's inside and outside. Just like the notion of territory, the concept of a border has various meanings and implications. It operates not only in political and geographical registers of sovereignty, authority, and jurisdiction, but also in legal, technical, and economic ones.[93]

The inside/outside dichotomy tends to neglect the dispersed nature of borders.[94] It also does not acknowledge that communities, identities, and political bodies are not restricted to either the inside or the outside of a state.[95] Instead, borders organize the relationships between inside and outside and redefine them in particular ways. For instance, borders can have the effect of creating new divisions and enlarge inequalities between entities, such as between people of different sexes or genders. Violence, in all its forms, is part of the daily practice of many migrants and related to borders and border controls in various ways.

Second, the erosion of the inside/outside distinction implies that various forms of registration, monitoring, and surveillance are materially and technologically dispersed. Dispersed borders express themselves in a variegated architecture of control that is spread over landscapes and seascapes spatially, which affects the execution of control at harbors, airports, and checkpoints.[96] Dispersed borders are likely to spread from the sphere of states and bureaucracies to households and private lives, even affecting the bodies of persons (e.g., by fingerprinting or facial recognition technology).

A third characteristic of Europe's border infrastructures is that the continent's materiality and spatial dispersal are entwined with knowledge infrastructures and information infrastructures to gather and process data— variously used to conduct risk assessments, support decision-making, profile migrants and travelers, identify critical border crossings, or detect patterns of mobility.[97] By tracking the trajectories of mobility, such as by gathering, storing, analyzing, and interpreting data, preemptive actions and preventions, such as no-fly lists, can be prepared with regard to persons far before they have reached an actual border.

Taking the inside/outside dichotomy as a starting point restricts the analysis of Europe's border infrastructures.[98] A too-dichotomous inside/outside distinction is likely to overlook the subtle filtering and selection process that borders facilitate and the coming into being of a multiplicity of classifications and categorizations of people on the move. Moreover, a too dichotomous inside/outside distinction might consider movement as the object of borders and borders as the objects of politics while their relationship might turn out to be much less subordinate. For that reason, this chapter has aimed to overcome the inside/outside dichotomy by following the rise of Europe's border infrastructures. The challenge is to arrive at an understanding of technological politics that allows issues of border control, mobility management, surveillance, and security to be articulated as a matter of politics.[99] If we follow this perspective, data clouds do not stand in opposition to barbed wire fences; automated border controls and facial recognition technologies at the airport have much in common with walls between countries. Attention to the morphology of politics requires an investigation of the mutual interaction between politics and technologies, between political ideas and the shape they take via borders. Political decision-making and technological border projects are intimately entangled. Border control technologies are vehicles for political thought. The resulting politics and policies can be seen as a conflictual world-making endeavor—one that constantly redefines the relationship between the inside and outside of Europe. The mosaic nature of surveillance at the airport, the patchwork nature of border surveillance technologies on land and at sea, and the appearance of movable humanitarian borders: all these phenomena are affected by the datafication of the border, while Big Data will likely further transform the relations between territory, mobility, and political subjectivity.[100]

The characterization of borders as infrastructures engaged with movement allows the beginning of a particular form of technopolitics. The following chapters of this book elaborate on how technopolitics functions as a vehicle for thought and action, particularly political thought and action. The management of mobility and border surveillance at airports, the creation of hotspots, and the European Union's borders in North Africa are not just things or policies that EU institutions and member-state representatives have decided upon, but technopolitical innovations of a particular political kind. As vehicles of decision-making about who is and is not allowed to enter Europe, they transport a particular political program, as well as the technologies attached to it. For that reason, an inquiry into the relationship between technology and politics is required.

Hungary rail track fence—closed September 2015.
Source: Henk Wildschut.

3 The Shape of Technopolitics

The Morphology of Borders

Europe's policies concerning international migration are an expression of the fraught struggle over sovereignty, jurisdiction, and mobility, revealing both state power and impotence. As constructions, borders materialize compromises between conflicting ideas and interests. Borders shape networks of circulation, instituting both crude and refined selection mechanisms to sort people. But alongside the deployment of barbed wire, ID systems, databases, and patrol boats, borders are bringing something else into motion: the machinery of governing, decision-making, risk assessment, and coordination. To study the variegated nature of borders and their entanglement with technology and politics, chapter 2 introduced the notion of "border infrastructure," which holds that borders are structures or networks concerned with the movement of migrants and travelers, as well as tools, instruments, information, experts, and knowledge that enable borders to classify and select people in order to manage human circulation. But borders and the technologies that comprise them are no mere instruments of political decision-making; they contain within themselves implicit or explicit political goals, generate unforeseen consequences, and encourage political intervention. Borders thus function as vehicles for politics. Studying their shape reveals the *morphology of politics*, as the materiality and spatiality of borders shape the technopolitics of movement.[1]

This chapter turns to the philosophies of Bruno Latour and Peter Sloterdijk, whose works point the way to a morphological notion of technopolitics, with which it becomes possible to navigate the materiality and movability of borders and border politics. The previous chapter described several characteristics of Europe's borders-as-infrastructures and introduced the notions of infrastructural imagination, competition, and compromise. But what kind of technopolitics do these characteristics give rise to? As I argued in chapter 1, there are many available options between technology as an instrument and technology as a way to view the world. Both Latour and Sloterdijk try to seize the *tertium datur*, the project of bringing the excluded third back in (namely the networks that connect objects and instruments to worldviews).[2] By reading Sloterdijk's theory of spheres through Latour's notion of actors and networks, I aim to develop an account of morphology and mediation that allows us to follow the movements between instruments, networks, and worldviews. As such, it deepens the concept of border infrastructures and relates it to the philosophy of technology and science and technology studies.[3]

Both Latour and Sloterdijk focus on the materiality and spatiality of the sociotechnical configurations in which technopolitics takes place. Their philosophies create awareness of the specific meanings of technopolitics and the various ways in which humans, institutions, and technologies are connected and confront each other. The aim of this chapter is to develop a morphological understanding of border infrastructures. This means transcending the boundary between politics and technology—and between ideas and things—by exploring how political ideas travel via things and technologies, as well as how devices, databases, and instruments are containers of political ideas and vehicles for political action. Based on the ontology, spherology, and political theology of Latour and Sloterdijk, the discussion here develops the concept of *peramorphic politics*—a morphological technopolitical account of how borders and politics give and receive shape.

My examination of Sloterdijk focuses on his *Spheres* trilogy, where discussion of artifacts, from buildings to ships and from the history of air conditioning to the use of poison gas in warfare, reveals an often-inspiring endeavor to conceptualize technology. Sloterdijk's conception of politics emphasizes the architectural spaces and places where politics is born. With its hyperbolic power, his work often kicks the spatial thinking of politics into overdrive by arguing that being itself is nothing other than extensiveness. But despite such metaphorical flourishes, I show in this chapter that Sloterdijk's repertoire is well suited for understanding how borders function as the vehicles of politics.

Latour's work is driven by an interest in mediation—in how humans, technologies, and nature intermingle and develop societies from the inside out. His work is concerned with the relationship between hybridization and solidification and how humans, ideas, and materialities group together to form social, technological, and political bodies through processes of association and translation and then disintegrate, redistribute, and reconnect. The following sections will show that a mediated account of movement, materiality, and space can contribute greatly to our understanding of the emergence of border infrastructures and the technopolitics that that entails.

Sloterdijk's Spheres: Bubbles, Globes, and Foams

Concepts of "technology" and "politics" are rampant throughout the literature on borders, explicitly and implicitly, but research into how technology and politics form, reform, and transform each other leaves much to be discovered. Precisely because a "border," both as a concept and as an entity, is mobile and to some degree indefinite, it offers a range of morphological modalities of technology and politics. The work of Peter Sloterdijk lends itself to explore these modalities well because he can be considered as a "morphological thinker" par excellence.[4] The space that we have as humans, and our possibilities for shaping our lives, create a common thread that runs through his work. The *enfant terrible* pursuing members of the Frankfurt School and their descendants investigates the position of the subject in an era when philosophical considerations have been overpowered by liberalism, economics, technology, and pragmatic thinking. This began with Sloterdijk's *Critique of Cynical Reason* (1988; originally 1983) and reached its controversial zenith with *Rules for the Human Zoo: A Response to the Letter on Humanism* (2009; originally 1999), in which he considered the biotechnological possibilities for managing the human "zoo" both now and in the future. The *Spheres* trilogy (Sloterdijk 2011, 2014, 2016; originally 1998, 1999, 2004) addresses the spaciousness of *being*—thinking, living, but above all building and housing—with the work of Heidegger frequently playing the role of whipping boy. According to Sloterdijk, being should not be thought of in relation to time, as Heidegger did in *Being and Time* (1927), but rather in relation to space. As chapter 1 explained, Heidegger (1977) argued that technology is usually defined in either an instrumental or an anthropological way. While the former places technology in a means-ends relationship, the latter regards it as the result of human activities. According to Heidegger, the essence of technology is not technological itself. Instead, technology contains a certain mode or attitude toward reality that unveils something about the relationship between humans and technologies, and ultimately about our being.

Sloterdijk takes a different tack. He situates being within the architecture of existence. According to him, being should not eclipse "becoming" and "moving," or how humans always create relationships with their surroundings.[5] Sloterdijk is interested in the locations of being, as well as in the spatial circumstances in which it comes into existence. It is, as he puts it, "a theory of humans as beings living in homes, and a theory of agglomeration of those beings in their diverse forms of living and gathering together."[6] According to Latour, Sloterdijk asks an architectural question, "one that is just as material as the geologists with their inquisitive hammer: where do you reside when you say that you have a 'global view' of the universe? How are you protected from annihilation? What do you see? Which air do you breathe? How are you warmed, clothed and fed? And if you can't fulfill those basic requirements of life, how is it that you still claim to talk about anything that is true and beautiful or that you occupy some higher moral ground?"[7] Meanwhile, these architectural and material issues are embedded in a theory of movement, as Sloterdijk explains in *Infinite Mobilization* (2020, originally 1989), in which he redefines modernization as a kinetic process.

Sloterdijk calls the architectures that result out of these movements spheres—environments of thinking and living that form climate zones within which the temperature can be regulated. These spheres not only are located in the subject's consciousness, but also are expressed in the buildings, the infrastructure, the means of transport, the media, and other technologies that we construct. Anyone who wants to understand being not only has to investigate the thinking "I," consciousness, or *Dasein*, but must also make the link to the cosmopolitan architecture of the spectacle-fixated consumer society within which it is shaped. For Sloterdijk, "Dasein is design."[8]

Sloterdijk's *Spheres* trilogy is very relevant to the discussion of border infrastructures. He reinterprets borders as creating immune systems. They separate the outer from the inner and, as with air conditioning in a car, create an agreeable climate as we race down the highway. This hyperbolic metaphor brings Sloterdijk to his understanding of politics: a formalized struggle for the redistribution of opportunities for comfort and psychological and physical well-being. It is a battle for access to the most favorable immune technology. From questions about the equitable distribution of goods, we move on to questions about the distribution of risks and opportunities for comfort and life chances.

Sloterdijk's conceptual repertoire, as developed in the *Spheres* trilogy, is underutilized in the analysis of borders. However, caution is required. A critical reading of the *Spheres* trilogy is needed to prevent an overly simple identification of spheres and immune systems with nations and national identities as "bodies," "organisms," or "homes." The notion of "immune systems" is telling in the context of migration politics and seems applicable beyond the comparison with walled states. The externalization of the border control policies of the European Union (EU) toward third countries, as described in the previous chapter, can be read as an attempt to regain sovereignty over the control of international mobility by introducing highly technological partnerships that displace the control of Europe's borders south of the Mediterranean, and increasingly south of the Sahara as well. According to Sloterdijk, this spatial expansion culminates in the process of globalization. To this end, in the first two parts of the trilogy, Sloterdijk examines the entire inner and outer space of Western cultural history. He does this in a literally breathtaking way, rewriting history as a climatological war—a struggle for the fresh air that is necessary to supply the spirit and the body with oxygen. The link that connects the inner and outer worlds of Western thought is represented in the first part of the trilogy, *Bubbles*, by the metaphor of the globe. Since the time of the Greek philosophers, the globe has stood for both the totality of beings and all that is good. The globe symbolizes the safe inner space (the womb, the home, the town), out of which thought emerged, as well as the wild outer space (the globe of the Earth), over which Western thought has spread, spiritually and materially.

Based on this metaphor of the globe, Sloterdijk examines the process of globalization in the second part of his trilogy (appropriately named *Globes*). Western intellectual history is reflected not only in material and spatial projects such as architecture and spatial planning, but also in conquests, crusades, discoveries, colonialism, exploitation, and globalization. In particular, *Globes* is concerned with borders, boundaries, and walls. The cities and walls in ancient Mesopotamia, Babylon, and China contained religious, military, and psychopolitical aspects. Some aspects of these ancient city walls still seem to apply to the present. Walls, according to Sloterdijk, cannot be fully grasped by their military pretensions, or by their claim to protect against enemies or outside forces. Walls also constitute a community or bring about a certain truth claim by establishing an unmistakable and self-explanatory entity. In that sense, walls are the material expression of the sovereignty of a group of people, a political unity—a sphere. "Spheropoiesis" is the notion that Sloterdijk uses to refer to this building of walls and creation of spheres. Spheres, however, are neither the direct translation of a state nor the will of the people into a bordered community. Sloterdijk is too much a media philosopher to rely on such an unmediated model.

Instead, spheres can be typified in the terms that Latour introduced. Spheres emerge via processes of *mediation* and *translation*: they arise in medias res. According to Sloterdijk, instead of being the expression of unified cities and communities or singular nation states, spheres are the product of co-isolation. Co-isolation should be understood not only in terms of the coupling of separated entities, but also as a prelude to the emergence of spheres. The process of bordering is not unlike how processes of coproduction are described in science and technology studies—namely, as the simultaneous development of political and technological. Co-isolation contains a moment of co-construction. Instead of regarding a border as a line dividing two existing entities, whether they are states, territories, or populations, the moment of creating a border can also be conceived as a foundational moment—or more precise, to use the term that was applied previously, a "mediating moment"—for the coming into being of co-isolated spheres.

From the outset, Sloterdijk claims, globalization has been driven by the pursuit of climate control, which is parasitic by nature. A climate has to nourish itself; it has to breathe and remain pure. Globes separate the outer from the inner and, like air conditioning in a car, create an agreeable climate while we race along the motorway guzzling fuel. The Western globe—the dominant one—is swallowing up the world. The "inner space of capitalism," as Sloterdijk calls it, is working overtime. Bringing in fresh air by turning up the air conditioning in our own economy inevitably leads to overexploitation elsewhere. The worldwide greenhouse effect, food and energy shortages, and the bursting of the soap bubble of financial capitalism (all of which are also due to rising temperatures) are the virtually inevitable results of this process. The huge flows of money, goods, information, and people have caused the Earth to overheat. The permanently high voltage of the continuously expanding Western comfort zone has made the indoor universe more pleasant but has left the outdoor space like a battlefield. The immune system is close to collapse.

In *Foams*, the final part of the trilogy, Sloterdijk examines more closely how these catastrophes are connected. He combines, somewhat dialectically, the microspherology of *Bubbles* with the macrospherology of *Globes*. It would be going too far to say that he is looking for a possible solution to the world's crises. But without saying it in so many words, he poses the question of whether and how the global spheres can still be collectively managed. *Foams*, according to Sloterdijk, offers "a theory of the present age from the perspective that 'life' unfolds multifocally, multiperspectivally, and heterarchically." It "can no longer be considered using the tools of ontological simplification."[9] To construct such a theory, Sloterdijk seeks to follow Latour's imperative to replace sociology with a theory of networks. Foam is the perfect metaphor for these networks, as:

> the lively thought-image of foam serves to recover the premetaphysical pluralism of world-inventions postmetaphysically. It helps us to enter the element of a manifold thought undeterred by the nihilistic pathos that involuntarily accompanied a reflection disappointed by monological metaphysics during the nineteenth and twentieth centuries. It explains once again what this liveliness is about: the statement "God is dead" is affirmed as the good news of the present day. One could reformulate it thus: "So the One Orb has imploded—now the foams are alive."[10]

The beginning of *Foams* is far from optimistic. Sloterdijk sets the tone by dating the beginning of the twentieth century in 1915, with the first gas attack by German troops in World War I. Where foam represents the fragile connection between isolated bubbles, the gas is the destroyer of this connection. The immune system is not only about to collapse due to overheating; it is also threatened by myriad new enemies.

The distinction between enemies of our own creation, such as the atomic bomb, and natural enemies such as viruses are not of great importance for understanding these immune disorders. Relentless developments in science and technology in particular are constantly improving our ability to identify and understand the threats to our biotope, and they teach us that the enemy is legion and cares little about our distinction between the natural and the social—witness the umpteenth announcement of a pandemic. The cell wall of the immune system is showing cracks; the inner and outer worlds are leaking into each other. Heidegger's homely dwelling, in which humans could unlock their true being, has become a playground for designers, engineers, lawyers, and politicians.

Sloterdijk's wordplay is often poetic and metaphorical. Sometimes this strengthens the expressiveness of his work; but at other moments, his hyperboles stand in the way of precision.[11] Nonetheless, we can find in Sloterdijk a philosophical vocabulary that offers important ideas for conceptualizing the workings of borders, the emergence of border infrastructures, and the nature of technopolitics. The biological and organic metaphors ("cells," "immune systems," etc.) are less important to me than his description of the coming into being and extension of architectures of thought, action, and movement. By constantly emphasizing the housing of ideas, Sloterdijk presses the history of European thought in a material-philosophical mold—a morphological reconstruction of the movement of ideas. Meanwhile, he describes how these thought-vehicles group together, organize movement, create tensions, culminate in wars, and allow the coupling and uncoupling of concepts. As such, Sloterdijk's spherology underlines two notions: (1) the morphology of technopolitics and the intimate relationship between thought and action, and (2) the movability of politics and technologies.

An example that is illustrative of the morphology of technopolitics and the movability of politics and technologies is the history of barbed wire. Barbed wire and the ideas behind it exemplify a particular aspect of modernity, one that connects violence, colonization, warfare, globalization, and—we might add—migration. Barbed wire was used to demarcate land ownership in the colonization of the American West and to prevent the enemy from crossing the lines in the battlefields of World War I. It shares a cruel history with the transportation, deportation, and encampment of people, as the concentration camps and the trains to Auschwitz testify.[12]

Barbed wire, razor wire, and concertina wire give the barriers, fences, and walls around the world a familiar look. The US-Mexico border, the Israeli West Bank barrier, the fence between Turkey and Syria, and the immigration detention centers in Australia and Libya are covered with their sharp edges and points, revealing the ambivalent relationship between borders and mobility. Concertina wire is a type of barbed or razor wire that comes in single, double, or crossed spiral coils. It can quickly be expanded like a concertina, an aerophone musical instrument similar to an accordion. Its applications include enhancing fences and walls and securing borders and military bases. According to Hebei Wanxiang Concertina Wire Company, a manufacturer based in China, concertina wire will "frighten or hurt" anyone who wishes to get through. As a mobile security barrier, it "is designed for rapid deployment" in "police and other special operations . . . in security and rescue or riot control." As a fence, it is a "very powerful device" to stop the "unwanted entry of enemies or animals." Its "sharp blades and spiral structure can trap anyone who intends to go through or over the concertina wire."[13]

Concertina wire can be found all over the globe. For many migrants, it is their encounter with the materiality of borders.[14] In 2005, thousands of migrants from sub-Saharan Africa were hurt by the sharp wires when they collectively tried to overcome the border fences of the Spanish enclaves Ceuta and Melilla on African territory. Concertina wire is also one of the main components of the border fence erected by Hungary on its borders with Serbia and Croatia in 2015 during the "migrant crisis" that was described in the previous chapter. The current fence will be equipped with other technologies, including devices capable of delivering electric shocks to unwanted migrants, heat sensors, cameras, and loudspeakers to blare messages in multiple languages. But concertina wire itself continues to be the defining feature of the border fence.[15] It has such symbolic significance that the German firm Mutanox refused to sell it to Hungary to be used for this purpose.[16]

Facts, Foams, and the Fabrication of Worlds

Two aspects of Sloterdijk's spherology that have been identified here (namely, the morphology of technopolitics and the intimate relationship between thought and action, as well as the movability of politics and technologies) clearly echo the voice of Latour. Not without reason, in *Foams*, Sloterdijk declares his indebtedness to his French colleague.[17] Suspicious of procedural forms of democratic politics and of human, all too human political constellations, Latour advances notions such as a thing being a gathering to underline the ontological entanglement of humans and nonhumans.[18] What does this say about different forms of politics? In *We Have Never Been Modern*, Latour (1993) explains that technologies in general and entities on the threshold of science and society, or nature and culture in particular, call for a special kind of political attention.

These questions evade the usual distinction between social and scientific problems that has maintained the division of labor between political representatives and scientific experts. Latour advances a precise definition of modernity, which he sees as a purification process. With his idea of a parliament of things, Latour suggests that the emancipation process that has brought modern democracy to humanity should now apply to objects as well.[19] In a way, such a parliament of the hybrid order had been installed several years earlier. Rather than the 1989 fall of the Berlin Wall, it is the Intergovernmental Panel for Climate Change (IPCC) installed in 1988 that typifies the rise of hybrids breaking the walls of modernist dichotomy. The modern era is made up of epistemological and ontological divisions where issues are taken apart and polarized, resulting in science versus politics, nature versus culture, subject versus object, and people versus things. But an ethnographic view of the history of modernity and its present-day manifestations (cities, infrastructure, laboratories, technoscience, and so on) enables Latour to argue that numerous interconnections have always formed a bridge between us and the environment, of which social and political groups have always been aware. The question is how these hybrids can be articulated in opposition to modernity's dichotomous ways of thinking.

Latour gradually abandoned his notion of a parliament. Indeed, the dissemination of issues and audiences and the distribution of power among public and private bodies indicate that the political is omnipresent, and not only in democratic institutions. Nor is the political only linked to representation and deliberation, the traditional forms of democratic expression. At the exhibition Making Things Public which Latour mounted with Peter Weibel in the Zentrum für Kunst und Medientechnologie in Karlsruhe in 2005, the visual played a central role. All these installations (it would be going too far to speak of "works of art") were examples of making visible the external effects of technology, which creates new audiences that become visible in public parliaments—meeting places outside the formal democratic order. Such parliaments, Latour argues, are urgently needed to give all hybrids that make up the technological globe a presence and place.[20]

In a contribution to the catalog for Latour and Weibel's exhibition, Sloterdijk endorses the view that a democratic system stands or falls on the possibility of catching political objects and making them public. For this reason, Sloterdijk in Foams talks of the need for "making the immune systems explicit."[21] Whereas Latour is concerned with a process of articulation, Sloterdijk talks of explication: the world is "explained" more or less as the IPCC has documented attacks on the Earth's biosphere—the same IPCC that Latour once classified as a parliament of the hybrid order that unites science and politics in a new way and gives a voice to both social and natural reality. These explication forms also apply to border infrastructures; as Sloterdijk says: "One of the hallmarks of progressive explication is that it expands the security arrangements of existence—from the antibody and dietary level to the welfare state and military apparatus—into formally secure institutions and disciplines."[22]

Like Latour, Sloterdijk sees little point in any strict demarcation between a registry of things and a registry of persons. In the twentieth century, it was science above all that supplied modernity. Science has transformed the modern world into a virtually uncontrolled, open-air experiment with the introduction of an unknown number of new entities, ranging from the double helix of deoxyribonucleic acid (DNA) to genetically modified organisms and nanoparticles. Even mathematicians share the blame for this, as evidenced by the invention of the notorious collateralized debt obligations (CDOs), the innovative banking products that were partly responsible for the financial crisis of 2007. Sloterdijk notes that Latour is a radical-democratic scientific optimist whose research is a joyful philosophy of a world populated with the products of explication. For Sloterdijk, Latour's contributions contain "the stimulus for an epistemological civil rights movement—with the aim of integrating the technical objects and animal symbionts into an expanded constitutional space, thus creating an integral republic that finally recognizes not only human agents, but also artifacts and creatures as ontologically fellow citizens."[23]

Affectionately, Sloterdijk calls his French colleague *Der Mann, der die Wissenschaften liebt* (The Man who Loves Science). Words of praise indeed, but reading between the lines, one cannot avoid the impression that he considers Latour to be a touch naive. What they have in common is their view of modernization as a kind of air freshener that tries to keep out the hybrids. Sloterdijk talks about the effects of the air freshener in terms of "immunization," while Latour talks of "purification." But whereas Latour is fascinated by all new entities that can find themselves a place in a shared socionatural order, Sloterdijk sees mainly hazards and enemies. Globalization creates one air bubble after another, from the Internet bubble and the millennium bug to filter bubbles on Facebook, Google, and Twitter and plunges us under a thick layer of foam. It leads to a worldwide bubble bath in which we are in danger of drowning.

The image of globalization as a worldwide bubble bath symbolizes modern pluralism. To Sloterdijk's great dissatisfaction, twentieth-century philosophy was able to think of pluralism only in terms of individualism. This began with Gottfried Leibniz's monadology, as how can there be a connection among atoms, cells, and people—in short, among all the individual "globes" in the cosmos—if these are all sealed off, windowless, and separate? Latour's solution lies in an extremely democratic ontology. As no other has done before or since, he has made an art of debunking all distinctions that are speculative rather than concrete in nature.

That relationship is not as distant as it seems. What Latour himself made clear early in *Irreductions*, part two of *The Pasteurization of France*, is that his research on science and technology is based on a specific theory of relations and objects. Latour's philosophy is deeply rooted in post-Kantian thought and debates in the late nineteenth and early twentieth centuries about the nature of reality and how it relates to our cognitive abilities. His philosophy bears unmistakable traces of William James (1842–1910), and especially Alfred North Whitehead (1861–1947). To Latour, this leads to a typical combination of realism and relationism. From this position, he arrives at a specific theory regarding what he calls "translations"—the way in which actors make connections between things that exist in reality, thereby changing and enlarging the world. This creates the networks that make up our world: irreducibly real and material structures that bring order to the world.

Of crucial importance when building these networks is the figure of actor as mediator. Anyone who wishes to give a political dimension to the role of the scientist will quickly recognize a Machiavellian aspect to this process. This is the case, except that this Machiavelli of Latour's is no Machiavellian. Latour reads *Il Principe* (The Prince) as a discourse on how institutions and relations are created if a new entity emerges and a protagonist appears on the stage to fill the vacuum. The researchers whom Latour studied, such as the chemist and biologist Louis Pasteur (1822–1895) and the physicist and chemist (and later Nobel laureate in chemistry) Frédéric Joliot-Curie (1900–1958), succeeded in doing medical or atomic research and adding new objects (molecules and microbes for Pasteur, atoms and neutrons for Joliot) to the world's ontology. But they were also able to relate the existence of microorganisms and microparticles to political and social issues like public health programs and military-strategic nuclear politics. This makes the actors, objects, and networks successful.

Like Latour, Sloterdijk falls back on relational ontology to escape the tragedy of the lonely particles. Sloterdijk's spheres theory is based on monadology, the theory of windowless particles by Leibniz. The notion of monadology is derived from the Greek word *monas*, meaning "unit." In Sloterdijk's case, monadology concerns a theory about how units such as nation-states unite and divide, how their borders travel to connect and disconnect, and how the units shape specific movable infrastructures. Like Latour, he makes the leap from the micro to the macro, the particular to the universal, the individual to the group. Sloterdijk's solution to the singularity that threatens to consign objects to isolation—thereby making politics, and certainly democracy, impossible—is to achieve a plural spherology. The problem of pluralism is redefined as a question of co-isolation. Sloterdijk's bubbles have small peepholes. In the digital era, these peepholes consist not only of intercom systems by the door, but the telephone, radio, television, and the Internet. Although it is true that each person exists in the bubble of their own living environment, these bubbles now all come together in a worldwide Twitter system. But how can one get from there to a political system?

Sloterdijk and Latour see politics as being closely related to the sociotechnical ordering of the world. Both embrace the idea that a political theory first needs a theory of society, but reject the idea that there is such a thing as society. Sloterdijk and Latour see the social as something fluid, a plasma, of which social institutions and social facts are the briefly coagulated results. Latour is not interested in *matters of fact* but in *matters of concern*: things that are in motion and create new links among people, technology, and the living environment. He calls the results of this process collectives or associations—temporary clusters of social and technical relationships that do not form a whole but are related to one another as networks. Facts, in this view, are temporary points of solidification that function as black boxes that contain all the information of their past and coming into being. This history of the coming into being of a fact is safely sealed and stored, and it starts circulating again only when a fact is disputed and reenters a controversy. The contestation of a fact, therefore, has much in common with opening a black box: all kinds of unexpected surprises that helped to keep the fact together but were safely forgotten to give the fact its face may show up again.

Latour underpins this view empirically with the results of his research. Conceptually, he plays an even bigger trump card: the fact that the social is fluid is nothing less than the reason for the existence of the political. If such a thing as a stable and well-defined society were to exist, which was knowable and representable, the political would represent nothing other than the status quo—and would thus abolish itself. Instead, Latour seeks a form of politics that can reflect that which can change: both the social and the political are looking for a form.

These concepts show similarities with what Sloterdijk calls "foam"—the lumpy collection of air bubbles of individual living environments that fill space but are empty inside. With his theory of foam, Sloterdijk wants to arrive at a new explication of social relationships. His example for the collective is neither the *Gemeinschaft* nor the *Gesellschaft*—neither the organic relationships of community nor the functional connections of society; he finds in political theory as little satisfaction in the communitarian society as in the liberal contract.

In their search for alternatives, Sloterdijk and Latour both look to the work of the French sociologist Gabriel Tarde (1843–1904). While Tarde's work has long been overshadowed by Max Weber and Emile Durkheim, Tarde has enjoyed a renaissance since 2000, not least due to the efforts of Latour himself, who is directly or indirectly responsible for a number of articles and books about him, as well as reissues and translations of his works.[24] Crucial for both Sloterdijk and Latour is Tarde's view that not only is a society made up of cells, but cells, molecules, and atoms (or any other smallest unit that one chooses) have a certain degree of organization and are small "societies." *Toute chose est une société* (Everything is a society). Instead of reductionism, Tarde already subscribed to relationism combined with realism. This radical relationism enables Sloterdijk to further elaborate his notion of co-isolation. It leads him to a theory of associations, in common with Latour as of late, with the proviso that for Sloterdijk, these associations do not so much create hybridity as spatiality. From the sociological definition, he arrives at a political definition.

In his search for a political alternative, Sloterdijk reviews three symbolic associations. The first is the national assembly, which he finds unsuitable for creating social synthesis.[25] The scale and complexity of the world's problems prevent such a localized model from being effective. He then turns to the stadium, which fits the image of a society focused on entertainment and mass spectacle. But the rhetoric of winners and losers that accompanies competitive sports impedes social integration. The image that appeals to him is the conference, where people sharing an interest gather in the short term, without forming a strict social community.[26] Conference-goers travel the world, temporarily connecting the worlds in which they live and think to create global associations. This idea of the conference resembles Latour's parliament of things, except that Sloterdijk's notion stresses even more strongly that formation, not representation, is what is at stake. Unlike a parliament, a conference is not a reflection of society, but forms it by bringing together the *res* and the *publica*.

The evaluation of national assemblies, stadiums, and conferences speaks to our concerns in this book, as Sloterdijk shifts our attention from how authority is justified in these gatherings to how they manage to combine various entities to emerge as unified beings, if only temporarily. Manifesting authority and expressing sovereignty are exactly what border infrastructures aim to do by bringing together all kinds of instruments, agencies, and information. Far from singular entities that speak univocally to travelers and migrants, border infrastructures are always in the process of creating their authority. Sloterdijk emphasizes that the "foam metaphor draws attention to the fact that there are no isolating means which are completely private property—one always shares at least one partition with an adjacent world-cell."[27] His three ideal types to represent and organize multiplicity also suggest a common concern with travel. To express authority, border infrastructures must travel, requiring all kinds of connections and collections. The movable politics of border infrastructures is grounded in the transportability of its composing elements.

Politics According to Sloterdijk and Latour

At this point, the question "What exactly is the nature of politics in a technological society?" can be asked again. The *Spheres* trilogy provides a provisional, and not entirely satisfactory answer. According to Sloterdijk, politics consists of the formalized struggle for the redistribution of opportunities for comfort and psychological and physical well-being. It is a battle for access to the most favorable immune technology. From questions about the equitable distribution of goods, we must move on to questions about the distribution of risk and opportunity for comfort.

At first glance, Sloterdijk's view on politics is an equally magical and metaphorical rephrase of what politics is about on the battlefields of the scorched earth where states aim to regain control over international migration. Still, something is strange about this definition. At its core, it is a formula in the style of Carl von Clausewitz: "War is the continuation of politics by other means." But if we take thinking from the inside out seriously and replace questions concerning our existence with questions concerning our extensiveness, does this not lead to the opposite conclusion? [28] Instead of claiming that the struggle for space, comfort, immunity, resources, and technology is a politics that continues by other means, it would be more consistent to say that politics is the continuation of spatial and material conflict.

This reading has consequences for how we conceptualize the politics of technology. Rather than seeing democratic politics as the forum of legitimate and collectively binding decision-making about the outside world, representation, deliberation, and legislation become the continuation of technologies by other means. In this sense, human democracy can be said to be contaminated by the ontology of technopolitics.

At times, Sloterdijk's rich use of metaphors and his unbreakable irony make it hard to distinguish playful provocation from rigorous philosophical analysis. Nonetheless, his reformulation of politics fits well in the aforementioned analyses of the material, nonhuman, and even posthuman conditions under which politics takes place, as well as of the hybrid gatherings where politics occurs. For his part, Latour falls short of a satisfactory answer to the question of how a democratic assembly or conference could be organized in which hybrids are represented, but strategic action is neutralized. It is odd that an author known for introducing Machiavellianism into the world of science and technology—"science is politics by other means," he once wrote—views politics in a way that has so little to do with power. Science, according to Latour, "does not reduce to power" but offers "other means."[29] While Latour describes the decisive role of actors in forging soft and hard networks and the stability and instability of institutions, he rarely considers inequalities of power as a problem. Whenever he appeals for more openness, less exclusivity, and political visibility for a greater range of problems, it seems to emanate from an idea of stakeholder democracy. While Latour describes this playfully as *Realpolitik* (because he is realistic about objects), his outlook has little regard for the power-conscious subtleties of political practices.

Latour's lack of attention to power and violence is exemplified in his description of a journey by the French marine officer and explorer Jean-Francois de La Pérouse (1741–1788).[30] La Pérouse was commissioned by Louis XVI to chart the Pacific Ocean. When he met Chinese people on what he called "Sakhalin," they entered into a discussion on whether Sakhalin was an island or a peninsula. To La Pérouse's surprise, writes Latour (1990) in "Drawing Things Together," the Chinese were not ignorant of cartography and drew the area on the map themselves. La Pérouse's task betrays clear colonial intentions and violence. His entire expedition disappeared under mysterious circumstances in 1788, while the guillotine would end the life of his client, Louis XVI, in 1793. None of this is to be found in Latour's writing.[31]

It is striking that Latour describes the relationship between La Pérouse and the Chinese population as "natural." In his usual ontological-technical way, Latour describes the events in terms of "inscriptions" and "calculation centers." Whether intentionally or not, Latour aligns himself with the neutral, technical tone of the French expedition. However, this description fails to recognize the history that enabled calculation centers to draw things together, or the power and violence that were often necessary for representations to circulate. Raw power has no place in Latour's descriptions. He thus remains close—too close—to the self-presentation of many scientific and technological projects (i.e., that of civilized and civilizing undertakings by gentlemen).[32]

How can we characterize Latour's and Sloterdijk's political philosophies? Rather than using a slogan such as "object-oriented politics," it is worthwhile to emphasize two characteristics: their focus on the material and nonhuman and their anti-individualism. Latour and Sloterdijk clearly do not adhere to the model of institutionalized liberal procedural democracy. Latour expands his program to show that politics, democracy, and the relationships that they manage must be accountable for the fact that we are part of a collective and hybrid problems are inescapable. In doing so, he broadens the sphere and scope of politics without indicating what other form of democratic politics would actually work. Likewise, it would be an understatement to say that Sloterdijk pays little attention to real, existing politics. His focus is on the material shape of ideas and ideologies and the ways in which they are made to travel, thereby underscoring that the content and form of politics cannot be separated. A further similarity between Sloterdijk and Latour is that both refrain from assuming a critical perspective on technology itself. Whereas Sloterdijk considers the heirs of the Frankfurt School as being too close to a humanism that does not grant objects and technologies the ontological and political status they deserve, Latour questions critical theory for too-often abiding by a naive realism, in which matters of fact are regarded as the ground of reality and the observer has a privileged position. This perspective overlooks the processual nature of reality and how humans and things may or may not end up in associations.[33]

Technopolitics breaks with the normative distinction between humans and nonhumans, as well as with the idea that technologies can be designed and regulated according to political will. Technopolitics tends toward a post-humanist view that acknowledges the deep entanglement between the agentic capacities of humans and nonhumans.[34] Building on these views, the study of states, borders, and infrastructures starts in the middle of things without having a view from above. But the mastery of technopolitics in the analyses of Latour and Sloterdijk does not mean that other forms of politics have ceased to exist. In addition to setting its own conditions for the form and shape of politics, the morphologization of technopolitics allows various forms of politics to enter onto the stage. The emphasis on ontological politics does not mean that the instrumental uses of apparatuses or the design of border environments according to political will has disappeared. Instead, the intimate entanglements among humans, institutions, and technologies that make up border infrastructures provoke all sorts of technopolitics to come into being.

Technopolitics as Peramorphic Politics

The remaining question is how Latour's and Sloterdijk's notions of politics can be understood in relation to border infrastructures. This final section develops the notion of "peramorphic politics" as a technopolitical, morphological account of borders. Rather than following Sloterdijk's organicist line of thinking and his emphasis on immune systems, my focus on the shape of technopolitics privileges the intimate entanglements that make up border infrastructures. This approach does not restrict itself to the study of the external characteristics of technopolitics (i.e., its material or technical extensiveness). It should be read as encouragement to consistently think inside out and outside in—an approach where technology is a vehicle for political thought and politics is a material endeavor.

For the purposes of this book, the important similarity between Latour's and Sloterdijk's views on politics is their focus on materiality, technology, space, and extensiveness. Both thinkers invite attention to the *morphology of technopolitics*. In *We Have Never Been Modern*, Latour (1993) proposes a parliament of things, a house that unites the chambers of nature and culture and gives space to the hybrids that shape the world. While this spatial and material orientation recedes in *Politics of Nature: How to Bring the Sciences into Democracy* (2004), it returns with a vengeance in *Making Things Public: Atmospheres of Democracy* (2005), a catalogue he co-edited with Peter Weibel. The interconnectedness of humans and nonhumans, technology and nature, and science and politics reaches its zenith in Latour's Gifford Lectures in 2013 and his subsequent book *Facing Gaia: Eight Lectures on the New Climate Regime* (2017). Building on the political theology of Carl Schmitt and Eric Voegelin, and engaging with anthropogenic climate change, Latour brings the concept of "globalization" back down to Earth. Again, he agrees with Sloterdijk, who (in Latour's words) maintains that "the complete oddity of Western philosophy, science, theology and politics is to have invested all its virtues in the figure of a Globe—with a capital G—without paying the slightest attention to how it could be built, sustained, maintained and inhabited."[35]

Thinking "globally" assumes a transcendental cosmology that allows a view from above—a God's-eye view from which all elements are related. In contrast, an Earthly perspective starts in medias res and acknowledges that there is neither a final cosmic order nor a final destination. This does not mean we are on our own, though. Latour argues, again following Sloterdijk, that the "globe is not what the world is made of, but a Platonic obsession transported into Christian theology and then loaded into political epistemology to provide a figure—but an impossible one—for the dream of total and complete knowledge."[36] Latour replaces the figure of the all-encompassing

Globe with one of the Earth. We Earthlings are earthbound—"'bound' as if bound by a spell, as well as 'bound' in the sense of heading somewhere, thereby designating the joint attempt to reach the Earth while being unable to escape from it."[37] This relationship with the Earth is not another blood and soil interpretation of Carl Schmitt's work, but a plea to pay attention to the composite human/nonhuman constellations of which we are a part and the impossibility of escaping from these entanglements via transcendental political theory. The same holds true for Latour's 2018 book on politics in the new climate regime, tellingly entitled *Down to Earth*.

Another contribution of Sloterdijk's trilogy is its consistent reading of institutions in material terms. The *Spheres* trilogy ends by mentioning the "city of foam"—a reference to the New Babylon of artist Constant (Nieuwenhuys) (1920–2005). For Sloterdijk, this project represents the capsular ontology of the lonely particles that connect and disconnect, the basic materials for building cities and forming conglomerations. Rather than considering parliaments as mere houses for political debate, his philosophy allows understanding political institutions and instruments as vehicles for thought that travel and expand through material expression and extension. A morphological account of technopolitics requires thinking from the inside out and replacing questions concerning our existence with questions concerning our extensiveness. Rather than framing the struggle for space, comfort, immunity, resources, and technology as politics by other means, it is more consistent to say that politics is the continuation of spatial and material conflict. Border infrastructures can be said to cast political ideas just as sculptures are cast from pouring wax into a mold.

Although neither Sloterdijk nor Latour advance what is commonly understood as political theory, their works provide building blocks to support the thesis that states are *laboratories of movement*. The "laboratory" here not only refers to the construction of large-scale infrastructures to gather information, compare situations, and control heterogeneous configurations; it also denotes the shifting in and out that is required to organize expeditions to *terrae incognitae*, as well as the circulation of flows that travel through it. The spatial and material notion of politics as developed in the works of Latour and Sloterdijk, I argue, is a valuable contribution to our understanding of the technopolitics of border infrastructures. The immanent perspective within this account of technopolitics allows studying the emergence of border infrastructures from the inside out. Rather than relying on a transcendental point of view from which border infrastructures can be normatively evaluated, it encourages us to study their inner mechanics.

To pursue such an investigation, I return to a concept introduced in chapter 1, but which can now be given more technopolitical shape: peramorphic politics. The account of technopolitics that Latour and Sloterdijk embrace is underpinned by the idea of immanence. The main characteristic of immanence that they elaborate is extensiveness—the multiplication of technopolitics via material means—which I argue is a fruitful way to understand borders. Borders mold politics and political entities in particular ways; the relationship between border control technologies and political thought and action points to the importance of forms and shapes. As I explained in chapter 1, border politics tends to be peramorphic, while the politics of mobility tends to be a kind of peramorphic politics, as the border is ubiquitous in the political thought on migration, mobility, and security.

One way to engage with Latour and Sloterdijk is to read their works as treatises on global governance, the emergence of networks of humans and technologies, and the ontological politics of mobility and security. The discussion of Sloterdijk's *Spheres* showed engagement with multiple spheres: biospheres, atmospheres, mental spheres, and public spheres. Sloterdijk's trilogy thus admits to multiple interpretations, of which I privilege two. The first holds that *Spheres* can be understood as an eco-organicist eschatology concerned with the final events in the history of the world and the destiny of humanity. This interpretation is particularly based on Sloterdijk's treatment of immune systems, climates, and comfort zones. A second reading is that *Spheres* is a morphological genealogy of globalization. By constantly emphasizing the housing of ideas, Sloterdijk presses the history of European thought in a material-philosophical mold, a morphological reconstruction of the movement of ideas. Throughout the trilogy, he describes how these thought-vehicles group together, organize movement, create tensions, culminate in wars, and allow the coupling and decoupling of concepts. Not without reason, the notion of "birth" plays a central role in Sloterdijk's oeuvre.

Seen through Latour's interpretation of the *eschaton* in his discussion of Gaia and climate change, the eschatological and generative readings of Sloterdijk's spherology appear to have much in common. The concept of peramorphology benefits from Latour's interpretation of Voegelin's work.[38]

Latour argues against catastrophic, apocalyptic, and eschatological views of climate change and for the continuation of time *after* the end of time. This argument can also be applied to border infrastructures. Rather than seeing borders as an *eschaton* in the sense of an end, a final act, they can also be seen as mechanisms, as entities that organize circulation and continue the process of movement *after* the act of bordering. The rejection of the eschaton as the final event also implicates the continuation of the border *after* the border. A peramorphic view seeks to elide final events by following Voegelin's imperative not to render the *eschaton* immanent (i.e., to not bring an order to its end, but to follow its continuation).

Reading Sloterdijk's spherology through Latour's actor-network theory brings out the mediating moments within the material movements that connect and disconnect actors, institutions, and technologies. Latour's notion of "black-boxing" holds that facts are temporary—provisional clots of knowledge, experiments, technology, and power that can always be opened and change within future network circulations. Actors, institutions, and technologies see all kinds of exchanges, delegations, and transformations that configure their mutual components and develop novel constellations. This relationship between fluidity and solidification, between continuation and mediation, is also reflected in the temporal and spatial modality of actors, institutions, and technologies. Rather than heralding the end of times, the notion of the *eschaton* can be seen as a mediating moment at which another transformation takes place. Analogously, this view can be applied to borders and bordering. As infrastructures that make selections and organize circulation, borders continue to mediate through the reproduction, transformation, and multiplication of the processes of selection and circulation.

This chapter worked with a morphological reading of technopolitics—one that focuses on the composition and extensiveness of border infrastructures. A morphological view of border infrastructures emphasizes two aspects in particular: their materiality and movability. Note that neither Latour nor Sloterdijk developed a detailed political theory; their contributions take the shape of a possible cosmopolitics (Latour) or spherology (Sloterdijk) with political implications. Their accounts make politics both more and less ubiquitous: less, because institutional forms of politics such as parliamentary debates, governmental decision-making, and international agreements are considered but one form of ordering the world; and more, because a greater number of practices, situations, and relations have the potential to become political within the technopolitics of border infrastructures. But arguably more important than politics becoming more or less omnipresent is its change in form. While political authority and jurisdiction may or may not shape the world, they are likely to operate in the mold of instrumental, material, architectural, and infrastructural configurations. The following chapters will unpack this idea by visiting border infrastructures in Europe, starting with an iconic hub of spheres and networks: the airport.

Fence, Ventimiglia, December 2016.
Source: Henk Wildschut.

4   Detection, Detention, and Design at the Airport

A Machine with Ever-Changing Wheels

At Schiphol International Airport's centenary in 2016, its architectural supervisor, Jan Benthem, made a clear case for air design.[1] "At the airport," he stated, "it is not about the outside, it is about the structures on the inside. It is a machine with ever-changing wheels, a transformation that the traveler ought to be aware of as little as possible."[2] Previously, in 1993, Benthem argued that "the biggest problem I faced while designing was that the terminal had to be a well-oiled machine on the one hand and a sort of living room where people felt at ease on the other, . . . creating a quiet, pleasant atmosphere where you don't see all the technology."[3]

The airport, one of the defining faces of the so-called global border regime, exemplifies a specific kind of technopolitics. It is one of the key locations where border control and migration management takes place. As such, airports and airlines play a crucial role in Europe's border infrastructures. Chapter 2 explained that border infrastructures (1) connect large-scale networks with local manifestations of borders, (2) select among migrants by organizing forms of circulation, (3) display a particular interplay between visibility and invisibility, and (4) can be movable entities themselves. At the airport, border infrastructures sort various kinds of migrants from ordinary travelers as they execute the visa policies of the European Union (EU) and act as gatekeepers. The result is that for a large part of the world population, traveling by airplane to Europe is impossible unless they have a specific visa, such as for temporal work, study, or tourism, or are able to apply for asylum. People that still seek to reach Europe, including refugees, are forced to take other, riskier routes over land and sea—chapters 5, 6, and 7 will describe this method of travel in greater detail. From a policy point of view, they then are regarded as "irregular migrants," a category that exists only because of the sealing of the airspace. Studying the role of the airport in shaping the European Union's global border machine and global mobility infrastructure promises to reveal the inner workings of technopolitics and the relationships among the various functions attributed to border infrastructures.[4]

The control over Europe's airspace emerges from various policies, including preclearance, carrier sanctions, visa policies, and coordination with buffer countries. Visa policies and blacklists of countries whose nationals require visas create a kind of "paper curtain."[5] Carrier sanctions support tight visa regimes. They require airlines to pay a fine for carrying aliens without valid passports or visas. All Schengen countries use these sanctions.

The role of airports in the technopolitics of Europe's border infrastructures can again be specified by applying the concept of mediation. Airports mediate in three related ways. First, there is internal mediation at the airport, the distinction, selection, and connection of various forms of circulation, including border control, passenger flow, and migration management. Second, airports mediate among existing national, EU, and international legislation and regulations, such as the visa regime and policies with regard to airline carriers and the operational task of the airport itself. Whereas the former concerns mediation at the airport itself, the latter concerns the mediating role of airports in Europe's border infrastructures at large. Third, the coordination between this internal and external mediation requires additional forms of mediation. To follow the imperative that was set out in chapter 3 (i.e., to detect the emergence of border infrastructures from the inside out), this chapter will start unfolding these forms of mediation by departing from the peramorphic mediations that take place at airports.

The notion of "peramorphic mediations" creates awareness for the transformation of borders and the ways that they connect and disconnect different spaces and situations. In some situations, borders act as objects or instruments of state power, other situations see borders functioning as networks (large-scale infrastructures to organize international human mobility), whereas borders also appear as a kind of worldview—a way of ordering reality. A technopolitical account of borders must acknowledge these transformations.

Three specific peramorphic mediations will be distinguished: design, detection, and detention. These mediations combine the related yet distinct functions of managing passenger flow, ensuring security, controlling the border, and managing migration. The dividing lines between these functions are highly porous, not least because the technologies to produce and reproduce this permeability remain in constant flux. But with the notion of peramorphic mediations, we will try to follow them.

The architectural supervisor at Schiphol's centenary touched upon the notion of mediation by implicitly pointing at three aspects—namely, materiality, spatiality, and visuality. Together, they create an infrastructure in which Schiphol's three spaces of design, detection, and detention relate to each other. At the airport, various dividing lines are at work, most notably the ones that distinguish the spaces of design, detection, and detention where the governance of regular (and sometimes irregular) travelers takes place. Rather than seeing infrastructure as the outcome of clear-cut political decisions and human design, it will be shown that the inner workings of the airport fuel a dynamic interplay among the spaces of design, detection, and detention, where events are made visible or invisible and become related or remain uncoupled.

This interplay is largely informed by the infrastructural compromises that mediate among the various tasks and services of the airport. The notion of "compromise" has a long tradition in politics and political theory, and this chapter explores one of its specific forms. The values and risks of political compromise can be approached from an ethical perspective.[6] Compromise can also be seen as an indispensable tool in democratic deliberations.[7] The approach that will be presented here is informed by a political theory of technology and aims to discover how the politics of the airport is organized internally.[8]

Infrastructural compromise in the context of borders and border management is concerned with the circulation and selection of people. As such, they mark moments where different movements are connected and come apart, at which some people are allowed to continue their journey and others are prevented from further movement. Instead of seeing compromise as morally acceptable or unacceptable—as a "second best" solution to specific problems or the outcome of negotiations between distinct actors—I conceive of compromise as an innovative means to bridge separate spheres. "A compromise suggests the possibility of a principle that can take judgments based on objects stemming from different worlds and make them compatible."[9]

Compromises promote "techniques of creativity" through which composite situations emerge and clashes are averted. The notion of "infrastructural compromise" helps us to understand the internal organization of the airport and how the machine with ever-changing wheels continues to select among different kinds of travelers in the political economy of international mobility. As such, infrastructural compromise echoes attempts in the governance of international mobility to present patchworks of information systems aimed at monitoring and control as a seamless web. Airports are infrastructural spaces in which risks of all kinds must be managed: from airplane security to baggage handling, from smuggling and migration to the prevention of terrorism.[10] This management is concerned with the circulation of all sorts of elements: passengers, luggage, goods, and information, but also risk and uncertainty. To analyze this governance of circulation, spatial and aesthetic perspectives can be combined.

Schiphol International Airport is an international hub.[11] Since 2000, Schiphol has had between 400,000 and 500,000 flight movements annually. Passenger numbers rose from around 40 million in 2000 to around 60 million in 2015, and over 70 million in 2019. Since 2015, it has ranked between tenth and fifteenth in the world in passenger traffic. But for our purposes, more important than its size and role in the global network is Schiphol's trademark concept of the AirportCity. Like Sky City in Hong Kong, Aviapolis in Helsinki, and Aerotropolis in Memphis, Tennessee, Schiphol's AirportCity choreographs the commercial logics of the "airside."[12] What makes Schiphol unique is that it has been selling this concept to other international airports since the 1990s. Not many studies of Schiphol International Airport combine a focus on the technical influence on policymaking with the social, more actor-oriented side of policymaking.[13] As a corrective, this chapter elaborates on the dynamics of the infrastructural compromises at Schiphol.

Airports are often seen as symbolic of the two faces of present-day border traffic.[14] On the one hand, the airport is an icon of globalization, embodying the free movement of people, money, goods, and information. On the other hand, the airport represents the restrictions that nation-states impose on the freedom of movement in a globalized world. In the opinion of numerous commentators, air passengers have become a "kinetic elite" that can move through the airport's clinical spaces unhindered by distance, space, or time.[15] Meanwhile, irregular travelers are dealt with behind the scenes so as not to inconvenience regular travelers. But this picture becomes more nuanced when the dichotomy of elite travelers versus irregular migrants or insiders versus outsiders is tested across different spaces. The hierarchical spatial order of the airport often remains hidden, not only to the regular traveler, but also in the literature that portrays the kinetic elite as the heroes of globalization and the airport as an icon of a life without boundaries.[16] For instance, the luxury network of airports, companies, and department stores has been termed the physical Internet—a seamless network where you can reach your destination with one click and buy whatever you wish with another click.[17] But a similar logic of design imbues parts of the airport where detection and detention take place as well. Encouraged by a governance style that embraces myriad pilot projects and the European homeland security market, the management of mobility, security, and surveillance is often portrayed to passengers as a service.[18] The treatment of different passengers and the relationships among design, detection, and detention reveals itself in a specific interplay between what becomes visible and what remains invisible once this service is delivered.

Considering the airport as infrastructure not only encompasses the large-scale technological networks of mobility, transport and security but the particular interplay between mechanisms of inclusion and exclusion, connectivity and collectivity, the visible and the invisible. By exploring the various peramorphic politics that connects the spaces of design, detection and detention at the airport, it will become possible to distinguish the various "infrastructural compromises" that establish these connections.

Design, Detection, and Detention

While various scholars have conceptualized the intimate relationships between design, detection and detention, how do they intermingle at a specific airport? Instead of appearing as a border checkpoint, border surveillance and the monitoring of passenger flows and migration management take place via processes of selecting and filtering persons.[19] These filtering and surveillance systems restricted to the location of the airport. They already start when people begin orienting on their journeys and buy tickets or arrange visa. The airport is deeply interwoven with urban life and society

as a whole.[20] Conversely, airports over the past decades have become cities in microcosm. While border control technologies may have penetrated
societies at large, societies now find expression at the airport. Design, detection and detention are intertwined at Amsterdam's Schiphol International
Airport.

Design: The Airport as a City

The peramorphic mediations of the airport have unfolded alongside evolving
architectural developments. In terms of its architectural history, the airport
originally reflected the railway station. In the early days of private air travel,
comfort left much to be desired. Although an air ticket was considerably more
expensive than a train ticket, the railways were able to provide passengers
with greater luxury. It was only when the military management of Schiphol
gave way to its operation by the City of Amsterdam in the run-up to the 1928
Olympic Games that serious attention was given to the facilities for travelers,
giving rise to a new "station building."[21] Nowadays, two different architectural traditions serve as examples: those of the city and the shopping mall.[22]

The imitation of the shopping mall involves practical requirements of
interior organization, as well as a specific program of logistics to enable
circulation—and more specifically, efficient circulation. Circulation, or
movement through space, not only concerns the logistics of architecture;
the concept has currency in the political economy of space more generally.[23]
The efficiency is not necessarily aimed at speed; it also needs to organize
dwell time. As such, circulation and the airport experience that it brings
forth is a compromise—one that unites the interests of airlines and shops.
It balances the relaxed speediness required by airlines so that passengers
can arrive on time for flight departures with the speedy relaxation required
by shops so that people can buy things on the way.[24] The resulting political economy is played out in terms of aesthetic considerations and design
largely concerned with "staging invisibility."[25] As such, the airport resembles the spheres that Peter Sloterdijk described. This can be illustrated with
the idea and the technology behind the shopping mall.

The inventor of the mall, the Vienna-born architect Victor Gruen, saw
it not just as part of the new city; the mall *was* the city. In the hermetically
sealed shopping mall, the enemy—the open window giving access to the
wild world outside—was neutralized, and the air deliberately cooled. The
shopping mall environment was conditioned by what came to be called,

very appropriately, "air conditioning." As the previous chapter showed, air conditioning is a central notion in Sloterdijk's *Spheres* trilogy. In the closing section of the second part of the trilogy, *Globes*, he predicts that air conditioning would be the space-political theme of the coming era.[26] Air conditioning as we know it today was developed in New York in 1902 by Willis Carrier, who called it "man-made weather."

The development of air conditioning and the ideas that came with it exemplify the struggle over comfort zones and life conditions as explained by Sloterdijk. The idea that not only buildings, but whole cities could be cooled had been floated as early as 1842 by John Gorrie, a physicist in Florida. Concerned about the effects of industrialization and urbanization on people's health and living environments, he became obsessed with the relationship between well-being and temperature. He even developed early machines that probably operated in a local hospital (but the sources are not clear on this). However, Gorrie was not striving to create a place with a pleasant indoor climate simply for humanitarian reasons. He was also interested in the commercial angle, with the idea that air conditioning could expand the world of trade. Air conditioning offered the possibility of creating the conditions for shopping in a purified atmosphere. Gruen saw the shopping mall as a society based on order and authority. In his view, a certain amount of planning was essential for this complex society. By creating the optimal conditions for physical and mental welfare, the mall protected nothing less than life itself. Protecting the mall from natural and social enemies guaranteed our freedoms.[27]

The history of the airport echoes Sloterdijk's *Spheres* trilogy. This line of thinking resonates with the renovation of Schiphol International Airport between 1963 and 1967 by the Dutch designer Kho Liang Ie. Kho "created a sort of three-dimensional background for flows of traffic and for travelers who were walking, waiting or resting . . . He made the spaces clear and unambiguous, balancing the busy, fast pace of travel by introducing calm, open spaces, light and long-lasting reliability. His Schiphol did not use much color; the passengers brought that aspect along with them."[28] Kho's design, however, received mixed reviews: "National Dutch newspaper *De Telegraaf* described it as 'science fiction,' while *Het Vrije Volk* considered the visitors' restaurant a 'room for suicides'; critic J. J. Vriend referred to a transport factory, and *Het Algemeen Handelsblad* missed a friendly, welcoming atmosphere." Some of Kho's inventions nevertheless became signatures, especially

the yellow signs on the ceiling. These signs guided passengers to their destinations. Other services, facilities, and all other nonflight information were
marked by green signs.[29] The use of glass and the installment of "lines of
sight" connected the inside to the outside and, while maintaining a clear
boundary, allowed passengers to view the planes waiting on the airstrip.[30]
Lines of sight generate "both a regime of perception and subjectivity and
a set of practices by which the lines of discrimination and partition are
concealed."[31] Inside the airport, time, space, and air were managed in such
a way that the airport-as-mall could flourish. As a result, the airport is to
globalization what the shopping mall is to the city—a comfort zone for
consumers who can have everything they want without having the feeling
of crossing a border or leaving its cozy, indoor space.[32]

The airport follows the logic of the city, but it also works in reverse—the
present-day city follows the logic of the airport. The mall, and in its wake
the airport, ended the idea of public space as developed in the nineteenth-
century European city.[33] With a historical center full of squares, cafés, and
public buildings, this kind of city symbolizes the public space of which dem-
onstrations, discussions, riots, and boisterous laughter—in short, noise and
activity—form an integral part. But the contemporary city has diminishing
regard for this classic, fundamental pattern. In an essay in *The Observer*, J.
G. Ballard (1997) states that at airports, we are no longer citizens with civic
duties, but rather passengers for whom all destinations can, in theory, be
reached. Following the system's rules, we travel light. According to Ballard,
airports have become discontinuous cities whose inhabitants are constantly
in transit but largely happy. Ballard likes airports because they show no trace
of kitsch or false nostalgia. He expects that the airport will become "a virtual metropolis whose faubourgs are named Heathrow, Kennedy, Charles
de Gaulle, Nagoya, a centripetal city whose population forever circles its
notional center, and will never need to gain access to its dark heart."[34]

Writers who saw the airport as a triumph over history welcomed Ballard's essay. But many missed the irony concealed in Ballard's words.[35] In
his work, the hard, gleaming outer wall of hypermodern architecture often
appears in contrast to the violence, lust, and anger that roar through the
city's veins. But whether this sense of irony exists for everybody is doubt-
ful. The architect Rem Koolhaas, for instance, enthuses about the only ideology that remains after postmodernism—that of the superlative. There

is no longer competition between styles, only between scale. Big, bigger, biggest—hence the title of the Koolhaas's illustrious book: *S, M, L, XL*. Koolhaas focuses on form and size, and he rarely comments on who should be responsible for the interior organization, strategy, and management of such a space. In this sense he embraces the description of the airport as a "non-place."[36] The image of a nonplace has been criticized repeatedly for obscuring "the complex pictures of power relationships that are enacted at airports through controlling, sorting and surveilling movement of people, things and data."[37] Instead of simply—and ideologically—reproducing the notion of a nonplace, Ballard creates an awareness of the intermingling of design with other practices.

Detection: Social Sorting and the Surveillance Mosaic

The experience of showing proof of identification at the checkpoint in exchange for access to the gates can be seen as a person transforming into an unchallengeable position as a passenger in the process of departing.[38] However, "in the post-9/11 era, "unchallengeable" is no longer an accurate description of the departing passenger's position. Rather, the passenger remains suspect, so long as they remain within or near a securitized airport, on an airplane, and on or near a tarmac."[39] Remaining suspect at the airport is a sociotechnical affair par excellence. Mobility management, customs and safety, security, and surveillance policies to prevent crimes, smuggling, terrorism, and illegal border crossings are carried out by myriad public and private professionals. Their work is structured by technological policies described in umbrella terms as "smart borders," such as the Schengen Information System (SIS) and the Visa Information System (VIS). Prevention on the spot is accomplished by using technologies such as body scanners, introduced after the al-Qaeda bombing attempt on Northwest Airlines Flight 253, an international passenger flight from Schiphol International Airport to Detroit Metropolitan Wayne County Airport, on Christmas Day 2009, and "a range of technologies, such as a [closed-caption television] surveillance system . . . ; iris-recognition and other types of biometric scanners; over a hundred metal detectors; fences; and a range of other devices."[40] More than 10 percent of all private security personnel in the Netherlands work at the airport—3,500 private security guards in the employ of the three leading security companies in the country (G4S, Securitas, and Trigion) and a host of smaller, specialized

companies. At Schiphol, "security is negotiated between all involved partners in a Platform ('Security and Public Safety Schiphol') that was set up as a
response to terrorism, of which the goal is to reach an 'integrated approach'
in which public and private partners make use of the same means and technology, each for its own specific goals and responsibilities."[41]

This continuous form of remaining suspect has often been described as
"social sorting." As theories of social sorting, classification, and categorization in the context of migration and mobility policy point out, inclusion/
exclusion is too bold a dichotomy to do justice to the nuanced groupings
and mappings that take place. Instead, a point of view is required that transcends a priori ontological opposition by emphasizing that individual persons and groups of people are assembled as a consequence of new policies
and technologies. The space of detection not only transgresses the practices
of border control, policing, and security, but also stretches the concept of
surveillance itself. The use of risk management technologies and the combination of biometrics with other databases, prescreening, profiling, and
dataveillance, along with increasing datafication, have led to a politics of
possibility and the performance of preemption.[42]

While the creation of insiders and outsiders and the selection processes
of social sorting, classification, and categorization have often been described
in ontological terms, they also contain aesthetic and spatial aspects. With
the concept of "the mosaic," it becomes possible to points to the "piecing
together" that takes place in security and surveillance policies "of other-
wise contingent life signatures." The resulting prophylactic profiles tend to
transform the space of detection: "the contemporary border is not merely
a site of technology where bodies become inscribed with code, but rather
it becomes the sovereign enactment of possibility."[43] This politics of possibility was exemplified in 2014 with the introduction of the first so-called
smart camera. The Royal Netherlands Marechaussee started a test at Schiphol
with smart cameras that detect abnormal behavior by travelers. This may
include clearly deviant behavior, such as wild arm gestures, but it also has
more subtle deviations, such as someone who leaves a suitcase behind. The
Marechaussee aimed to use this technology to trace criminals and prevent
attacks.[44]

The presence of approximately 3,500 cameras and the aim of creating a seamless passenger flow make Schiphol an ideal test lab for artificial

intelligence applications. In 2018, the TRESSPASS project started at Schiphol. TRESSPASS elaborates on the idea of risk-based security checks. It offers a framework for modeling risk, as well as a systematic approach of quantifying risk. TRESSPASS is based on a set of indicators that can be measured across all tiers of the Integrated Border Management program.[45] At Schiphol, TRESSPASS applies deep learning instruments. According to the director of VicarVision, one of the companies involved in the project, instead of using facial recognition technologies, and in accordance with Louise Amoore's notion of the mosaic, TRESSPASS interprets a person as a "collection of pixels."[46]

The notion of the "mosaic" is likewise applicable to the apparatus that composes these risk assessments and pictures of potentiality, as well as to the organization and the interior design of the airport itself. Far from being a nonplace, the airport is a space of potentiality, where an aesthetic ontological politics of possibility is at stake—one that is intimately connected to the program of design and the architectural and technological development of Schiphol. At Schiphol, the so-called trusted travelers use the Privium service program—a public-private partnership of the Schiphol Group and the Immigration and Naturalization Service, which allows these travelers, for a fee, to use an iris scan to accelerate their crossing of the border. The Privium Club Lounge is reserved for passengers to make their wait more comfortable.[47] Irregular migrants are led through various locations, of which the detention center is the most iconic, as we will see next. But the detention center, too, is a less singular entity than it appears at first sight, and it is better understood through its composite parts and visual representations.

Detention: The Space of Exception

The detention center has come to exemplify contemporary forms of exclusion. In the short film *Seamless Transitions*, the British artist and technologist James Bridle explores three architectural spaces that are part of the immigration system of the United Kingdom (UK). One of them is the Inflite Jet Centre in Stansted Airport. The center houses people who are to be repatriated because they have not been granted asylum and have now reached the end of that process. The deportation center is unphotographable for reasons of security, secrecy, or law. So how did Bridle manage to portray it? According to Bridle, he "had to acquire planning documents and satellite photos, interview academics and activists, and read the reports and accounts of those subject to their machinations. Working with Picture Plane, an architectural visualization firm, we recreated the three spaces as [three-dimensional] computer models."[48]

Schiphol's detention center is at the airport's original location, an area now called Schiphol East. When people are judged not to have a valid right to stay and are most likely to be sent back to where they came from, they will be held as irregular migrants. Considered so-called illegals, they will spend their time before deportation in a temporary detention center (usually the cellblock at Schiphol East). To deal with these cases, there is a judicial complex with a court, a district public prosecutor's office, an office of the Royal Netherlands Marechaussee, and a cellblock that consists of a detention center and a deportation center. These centers were made possible by the Drug Smuggling (Emergency Measures) Act of 2002 and were initially intended only for drug couriers, known as "mules." Illegality per se is not a punishable offense under the criminal code; rather, it is an administrative offense. Nevertheless, illegals are also detained in these cells. The fact that these people have not committed any punishable offense does not afford them any advantages. In these centers, there are several people in each cell, even sometimes families with children. The facilities are generally worse than those in regular prisons, and the regime is stricter; access to doctors and lawyers is limited and there are fewer possibilities for visits or psychological help. In contrast to other prisoners, foreigners in detention are not allowed to work and have no right to either training or education. Boredom is widespread, but complaints are few; there is a widespread fear of saying the wrong thing, which could seal one's fate.[49] The detention, which can run to several months, is no fun. For comparison, to be placed behind bars for longer than six months in the Netherlands generally requires a serious crime.[50]

Detention is a specific and extreme example of the visible-invisible distinction. What was hidden from the public eye became national news on the night of October 27, 2005, when a fire broke out in the detention center in Schiphol East. Eleven detainees were unable to leave their cells before the fire reached them, and they perished. Apart from the event itself, two specific aspects of the news coverage and the subsequent investigation into the causes of the fire add a further layer to our understanding of the interplay between the visible and the invisible.

First, the notion of "illegals" appeared troublesome, not only because illegality is a slippery legal and administrative category, but also because the news items initially paid more attention to the fact that three illegals who had tried to escape were arrested in the vicinity of the complex over the course of the morning than to the people who died.[51]

Second, the investigation into the causes of the fire became a political lightning rod. The investigation by the Dutch Safety Board concluded that "the Detention Centre Schiphol-Oost was insufficiently prepared and set up for an outbreak of fire"; that both "the main directorate of the DJI [the part of the ministry responsible for detention centers] responsible and the Site Manager were insufficiently critical in their assessment of the fire safety"; and that "the Municipal Council of Haarlemmermeer [formally, Schiphol is part of the Municipality of Haarlemmermeer, not Amsterdam] discharged its role insufficiently."[52] The investigation also produced a meticulous visual reconstruction of the event that became infamous for its dramatic musical accompaniment and its accusatory style. What is interesting from the point of view of aesthetic ontology is that the visual reconstruction reproduced the material, spatial, and visual repertoires that shaped the center in the first place.

In this light, the airport is anything but a blank space. It is not a nonplace where citizens become passengers, a place where urbanism swings free of social, cultural, and historical structures. A highly specific aesthetic ontology sets the tone and determines the interior organization of the airport. In a way, the detention center places migrants both legally and physically outside the normal order, where the limits of the law are tested and sometimes transgressed. This speaks to the argument that Europe's migration policies create "states of exception," of which the migrants who are detained in centers are an example.[53] Architectural projects, such as by Koolhaas, can create smoothly operating zones for living, working, traveling, and shopping only if everything disagreeable to the urban consumer is excluded. In that sense, the detention center is the counterweight to the sealed areas of the airport, but also of other cocoonlike places like shopping boulevards, urban promenades, theme parks, and shopping malls.[54] In the era of globalization, the airport further advances what shopping has done to the city: open spaces must be viewed with suspicion and replaced by a soothing, comfortable climate sealed off from unpredictable forces. Although the "states of exception" argument is powerful, it is also important to underline that the several spaces that regard migration policy are not entirely separate. The previous analysis suggests that there are much closer ties between what is considered open and closed or visible and invisible.

The following chapters, which analyze the hotspot approaches in Greece during the migrant crisis of 2014–2016, will also explain that registration and detention centers may create states of exception, but they are part of a network of policies, institutions, and technologies; and they form a very mobile border infrastructure. As at the airport, as explained here, the various spaces of migration and border policies connect and disconnect and are part of a kind of border carousel.

Infrastructural Compromises and Aesthetic Politics

The analysis of the distinct yet related spaces of detection, detention, and design resonates with the infrastructural characteristics of the airport. Schiphol International Airport houses both arms of government and private companies. Guarding and monitoring the border are the responsibility of the state, carried out by customs and the Royal Netherlands Marechaussee. But the infrastructural space is run by the Royal Schiphol Group, a company that also manages the airports of Rotterdam and Lelystad, as well as Terminal 4 of John F. Kennedy Airport in New York. JFK International Air Terminal LLC (JFKIAT) is the operator of Terminal 4 at John F. Kennedy International Airport. Founded in 1997, JFKIAT is owned by Schiphol USA Inc., a US affiliate of Amsterdam based Royal Schiphol Group.[55] At Schiphol, distinctions are drawn between those who have a right to stay in the country, for shorter or longer periods, and those who have no such right. But the process of social sorting does not function as a simple sieve; it rather composes mosaic pictures. Composing these pictures leads to overlapping functions and spaces, as well as collaboration between myriad public and private professionals concerned with mobility, security, and surveillance. As such, there is an intimate relationship between restricted and unrestricted spaces, strategies of inclusion and exclusion, tasks performed by government bodies and private firms, and the Crown Lounge at Schiphol and the euphemistically named *grenshospitium* (border hostel) at Schiphol East. The airport thus distinguishes between restricted and unrestricted spaces, between places where traveling should be a seamless and immaculate experience and places where the border has a strong physical presence.

The unstable relationship between what is visible and invisible, accessible and inaccessible, is defined via interactions between various requirements and aesthetic ontologies. The airport has myriad spaces that lack an overarching principle. The airport's technological organization can be understood only from the inside out. The planning and daily workings of large infrastructures depend on interventions by all kinds of human and nonhuman actors that relate the macro to the micro scale and vice versa.[56] The absence of a bird's-eye view has important sociotheoretical and political implications. One of these is that every perspective is constructed because there is no natural vantage point. Another is that without a view from above, no map can claim to be complete. Contrary to what is often claimed for the panopticon, surveillance does not take place from a single vantage point. Instead, it combines all kinds of local and regional networks via interoperability. The circulating information is then carefully reconstructed into representations that create situational awareness, which call or do not call for action.

The most obvious way to proceed from here would be to claim that the airport has multiple ontologies, with the different (though related) practices of design, detection, and detention being driven by different material, spatial, and visual infrastructural logics. Although the statement holds true to a certain extent, a more precise way to grasp the underlying relationships among these spaces is to identify the different peramorphic mediations at work. Design at the airport is not just a cover-up operation to mask processes of detection and detention. Detention is not just the extreme outcome of a selection process. Instead, design, detection, and detention generate different processes of mediation. The compromises that must be reached to facilitate passenger flow management, security policies, border control, and migration management, to combine security policies with service polices, speediness with relaxation, open spaces with closed ones—all of these point to the coming into being of various infrastructural spaces.

At this point, we need to delve deeper into the notion of "compromise." A compromise is a way to bridge tensions among different worlds. A compromise can be a regulation or an institution; for instance, paying attention to workers' rights can be seen as a compromise between the industrial and civic worlds.[57] But can compromises also express themselves in infrastructural innovations? Wildlife crossings such as underpass tunnels, viaducts,

fish ladders, and amphibian tunnels can be seen as infrastructural compromises between economic considerations of mobility and ecological
considerations of keeping habitats connected. The Eastern Scheldt storm
surge barrier in the Netherlands is an infrastructural compromise that uses
sluicegate-type doors, which allow saltwater marine life and local fishing
behind the dam but can be closed when weather conditions require it.

Like the storm surge barrier, doors play a mediating role at the airport,
physically as well as metaphorically. Revolving doors in particular exemplify a compromise in the governance of the AirportCity—a compromise
between the control of the circulation of people and the regulation of the
inner climate of the airport—revolving doors ease interior and exterior air
pressure differences and help regulate the climate in air-conditioned buildings; the first patent was granted to the *Tür ohne Luftzug*, or "door without
draft (of air)" in 1881. Revolving doors also serve as a metaphor for how
areas of the airport relate to each other. The spaces of design, detection, and
detention all set different kinds of borders that control the movement of
people. But how are these spaces mutually organized? The metaphor of the
revolving doors suggests that there is no transcending organizing principle
at work, but rather an immanent logic of bordering that connects as well as
separates the three spaces. The atmospheres of detection, design, and detention become "co-isolated," as Sloterdijk would describe it.

The emerging spaces do not demarcate opposite functions, but rather
follow an aesthetic movement in the sense of a specific political interplay
between visible and invisible. The aesthetic politics of the airport allows a
specific distribution of the sensible.[58] Through these aesthetic processes and
compromises, spaces are connected and unconnected. Rather than being
functionalist domains, design, detection, and detention designate specific
spaces where a certain reach is exercised.[59] The infrastructural constellation
of Schiphol International Airport comes close to that of an interstructure—a
concept and an entity that receives shape while circulating within and
through the peramorphic politics of the AirportCity.

Transformative Technopolitics

This chapter has examined infrastructural compromises in the context of
mobility, security, and surveillance at the AirportCity through an aesthetic
and spatial ontological lens. Infrastructural compromises were examined
in the three distinct but related spaces of design, detection, and detention

where passenger flow management, security policies, border control, and migration management take place. The analysis suggested that the notion of infrastructural compromises can shed light on the specific interplay between the visible and the invisible at the airport. Instead of being clear-cut dichotomies, what is open and what remains closed are intimately related and often appear in the form of compromises, combined actions, and composite pictures. Security and service, speed and relaxation, regular travelers and irregular migrants, and suspects of legal and administrative offenses are materially, spatially, and visually divided and connected. Far from being a nonplace, the airport allows people, goods, capital, and information to circulate while offering spaces in which innovative compromises can arise—infrastructural compromises that peramorphically connect the distinct spheres of the airport in order to support its internal governance within the political economy of international mobility.

The notion of peramorphic mediation emphasizes that technological networks—be they small or large scale, local or transnational—are constructions and compositions; they not only build on existing spatial-temporal-material conditions, but themselves shape novel infrastructures. Notions such as "European" or "global" all too easily assume a preexisting structure or relatedness that must be created first. Border control technologies and infrastructures here are no exception. Notions such as "the global border regime" all too easily assume that spaces of circulation are somewhere out there, spread out across the globe, waiting to be visited. However, this portrayal is misleading.[60] The search for the technopolitics of border infrastructures must pay attention to the composition of specific border configurations and the differences among various locations and spaces where control and selection take place.

The study of Schiphol International Airport shows that the construction of a global border regime requires all kinds of local, spatial, and architectural arrangements in order for airports to function as gatekeepers in international mobility. The technopolitics of border infrastructures at the airport can be conceived as a configuration of infrastructural compromises to bring together various functions of the border and to regulate different aspects of mobility. As such, airports are crucial centers of coordination and circulation in international mobility.

It might seem a big step to go from airports to border control and migration management on the Greek Aegean Islands. The next two chapters will describe the efforts of the European Union to execute border surveillance in the Aegean during the so-called migrant crisis of 2014–2016, and how border infrastructures to some extent traveled with the migrants who arrived there. Although it is hard to imagine a larger gulf separating the designed spaces of Schiphol and the reception and registration centers in Greece, between the luxuries of international flights and the dangerous and humiliating crossing of the sea from Turkey to Greece with the help of expensive and often unreliable smugglers, both passages and passage points are part of contemporary border infrastructures. Moreover, as the introduction to this chapter emphasized, airports are directly related to border crossings on land and at sea, as the visa regimes of European states executed by airlines do not allow certain nationals to board without meeting additional requirements, thus forcing them to take other, often more dangerous, routes.

Still, it is a leap from Schiphol International Airport to the islands of Lesbos and Chios, to which we turn in the next several chapters. Although the management of passenger flows may seem a far cry from the management of international migration, there are some underlying similarities. In a sense, the international coordination of the monitoring of human mobility is not unlike the monitoring of air traffic. But of particular interest, for the purposes of this book, is the workings of technopolitics. Like airports, European borderlands and borderseas are increasingly seen as spaces that should be brought under control. The "crisis" jargon that peaked in 2014–2016 and the extraordinary measures taken in its wake undoubtedly reflect a sense of taking back control. The particular forms of border control practiced in Europe today reveal the political preoccupation with governing international mobility. Like the construction of a global border regime, border surveillance in Europe on land and sea aims to collect, connect, and coordinate information about human mobility to attain an overview, although border infrastructures are far from a seamless web in practice. In the meantime, border infrastructures are turning local and regional places into areas under surveillance, a process perhaps most prominent in the European Union's hotspots. And needless to say, the moving, expanding border affects people and organizations, whether they be migrants, state agents, nongovernmental organizations (NGOs), or local volunteers.

To detect the specific technopolitics of border infrastructures at work on land and at sea, the next few chapters will again approach these infrastructures from the inside out. They will pay attention to the genealogy of technologies and the politics that have encouraged their development, but they will also examine border infrastructures—configurations of people, politics, and technologies—as moveable entities. The emphasis on movement also harbors a conceptual issue: namely, that the dichotomy of human mobility/border stability is misleading. Instead, these chapters will sketch the compromises among international mobility, border infrastructures, and politics, in which the constituting parts not only respond to each other, but tend to travel together as well.

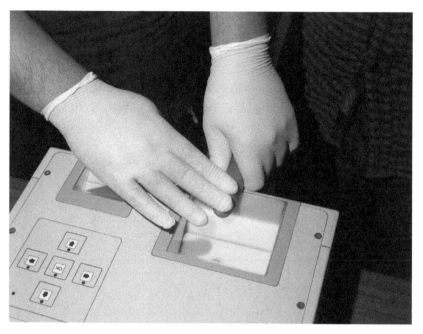

Fingerprinting in Hungary, May 2015.
Source: Henk Wildschut.

5 Surveilling Landscapes and Seascapes

Compromises on the Greek Aegean Islands

If borders can be considered as mobile infrastructures and as vehicles of politics that transport political ideas and actions, then they can also be regarded as entities that create technopolitical maneuvering spaces. The emergence of infrastructural compromises, as discussed in chapter 4, indicates the presence of such a maneuvering space. Technopolitical maneuvering spaces harbor both political intentions and technological instruments. Whereas the previous chapter discussed the peramorphic mediations among detection, design, and detention at the airport, the peramorphic mediations that figure in this chapter and the next one concern the connections among migration, security, and humanitarian aid at the Greek Aegean Islands during the period of 2014–2016.

The focus here will be on the northern Aegean Islands of Lesbos and Chios. Lesbos, the third-largest island in Greece, is only 5.5 kilometers from Turkey at the narrowest point of the Mytilini Strait. Chios, the fifth-largest Greek island, is situated 7 kilometers from the Turkish coast. The complicated relations among border surveillance, search-and-rescue operations at sea, and the creation of hotspots (which will be discussed in this chapter), and between care and control and the rise of a "humanitarian border" (which will be discussed in chapter 6) can be regarded as the result of an interplay between conflicting yet related concerns. The chapter will also attend to compromises between border infrastructures and various landscapes and seascapes as border control intermingles with different terrains and territories, which affects the relation between vision and action.

The period of 2014–2016 was a landmark in the development of border infrastructures in Europe. The emergence of border infrastructures and their movement across landscapes and seascapes is inextricably linked to the growing number of migrants who have entered Europe since 2011, especially from 2014 to 2016. Although many of these border control initiatives grew out of existing programs, systems, and policies—and could even be said to accord with the initial ideas underlying the Schengen Agreement—the "migrant crisis," as it is often called, intensified the technopolitics of Europe's borders. On October 27, 2015, Donald Tusk, the president of the European Council, stated in his address to the European Parliament:

> The crisis, or rather challenge, that we, all of us, as a community, as the European Union, are facing now is perhaps the biggest challenge we have seen for decades. I have no doubt that this challenge has the potential to change the European Union we have built. It has the potential even to destroy achievements such as border-free travel between Schengen countries. And what is even more dangerous, it has the potential to create tectonic changes in the European political landscape. And these are not changing for the better. These are truly extraordinary times that require extraordinary measures, extraordinary sacrifices, and extraordinary solidarity.[1]

The "extraordinary times that require extraordinary measures" largely happened in Greece. Due to the war in Syria, conflicts following the Arab Spring, and the merging of migration routes from various countries in Africa, Asia, and the Middle East, Greece—according to the International Organization for Migration (IOM 2017a)—received 1,047,939 undocumented migrants by sea in 2014–2016, which represents about 66 percent of the 1,582,759 arrivals by sea recorded in all of Mediterranean Europe during this period.[2] The Aegean Sea between Turkey and the Greek Aegean Islands became the main entry point to Europe following the closing of the land route between Turkey and Greece. In 2011 alone, around 55,000 migrants were detected crossing the Evros River on the Turkish-Greek border. The Greek government's completion of the controversial Evros fence in December 2012 encouraged many migrants to opt for the overseas route. IOM (2017a) states that the flow began to increase in June 2014 (with 6,214 recorded arrivals) and continued to grow until October 2014 (11,628 arrivals). In October 2015, it reached 217,936 persons—that is, more than 7,000 persons on average per day. Arrivals by sea to Greece decreased significantly after efforts to close the so-called Balkan Route by Hungary and the subsequent deal (known as the "Statement") between the European Union (EU) and Turkey.

This chapter pays particular attention to the emergence of the European Surveillance System (EUROSUR) and the European Union's "hotspot" approach. EUROSUR, which became operational in December 2013, is one of the defining programs of European border technopolitics and remains active on Lesbos and Chios and their surrounding waters. The hotspot approach identified the Aegean Islands as experiencing "disproportionate migratory pressure"[3] and intervened accordingly. The analysis will show the connections between EUROSUR and the hotspot approach and describe how peramorphic mediations between border control technologies reshaped the way that borderlands and borderseas are perceived and represented, as well as how it results in a particular relationship among humans, technologies, and the terrain where borders are drawn.

Chapter 4 explained how the internal organization of a specific border configuration (namely, the airport) could be understood as a series of compromises between the spaces of design, detection, and detention. But how do the technopolitics of borders work out when compromises must involve multiple states and organizations spreading across Europe? Compromises are not restricted to negotiating actors like persons, organizations, or state representatives that express their conflicting interests, plans, wishes, and desires. Compromises do not have to be limited to interactions in which the parties involved give and take a little in order to reach an agreement. The notion of a compromise can also be used to point at the innovations that take place to connect various ideas, practices, and techniques. Compromises emerge out of mediations. In such cases, a compromise is a kind of bridge that not only connects two sides, but also installs a new entity.

An effort to arrive at such a bridge is the 2015 European Agenda on Migration (discussed in chapter 2) and the subsequent policies, such as the deployment of the EUROSUR program and the hotspot approach. The argument that will be made here does not say that the EUROSUR program and the hotspot approach originated from the same policy agenda or have a common political source. Instead, it states that the similarity between EUROSUR and the hotspot approach is programmatic. Both contribute to the intermingling of monitoring on land and at sea, as well as to the connection between visuality and action. Further, although the two have different genealogies and were indeed developed separately, the joint mention in the press release is meaningful, in that it precedes the policies that were taken during the migrant crisis, as expressed in the important 2015 European Agenda on Migration, which connected EUROSUR and the hotspot approach. This document expanded border security as it referred to border protection, as well as the duty to prevent the further loss of life by migrants and refugees.[4] Saving lives and securing borders became the two key security imperatives. This compromise between security and humanitarianism ought not to be seen solely as a covenant between actors or concepts. A technopolitical compromise is not a contract between citizens and migrants; instead, it is a very material entity.[5]

What are the consequences when this argument is applied to the compromise between security and humanitarianism during the migrant crisis? First and foremost, it means that the compromise ought not to be analyzed solely at the level of policy paradigms or political ideas. Political ideas concerning the security identity of the European Union and the common task to control the external borders on the one hand, and the will to offer humanitarian care and to prevent violations of fundamental rights on the other, are expressed in a particular infrastructural space that offers opportunities for, but also poses restrictions on, the realization of actions and ideas. Conversely, compromises should not be reduced to the creation of concrete objects. Identity cards, for instance, allow people to be registered for border security reasons, but also support the exchange of medical files to provide health care. In a material way, identity cards are a compromise between security and humanitarian aid. This material realization of objects is an important focal point of a morphological view on border infrastructures, but it has to be related to the development and dissemination of political ideas.

To accomplish this, this chapter and chapter 6 will discuss the technopolitical compromises that have generated the European border surveillance system EUROSUR, the hotspot approach, and the humanitarian border on the Greek Aegean Islands. This chapter considers the surveillance of landscapes and seascapes, paying specific attention to the intermingling of various geographies and territories via the construction of compromises. The tensions, compromises, and interactions consist of the merging of surveilling practices on land and sea and a specific intermingling of seeing and acting (i.e., of visualizing events such as the entering of unregistered vessels in territorial waters and intervening in these occurrences). But instead of leading to an all-encompassing surveillance regime, intensified border control and migration management policies developed into a kind of border bricolage of agencies, policies, institutions, technologies, and geographies.

The analysis in this chapter is partly based on a series of interviews that were conducted at Athens, on Chios, and on Lesbos during the period of

2014–2016. The interviews were geared toward identifying the surveillance practices that were related to EUROSUR and the hotspot approach. Excerpts of these interviews are employed here to illustrate the development of surveillance technologies on land and at sea and the tensions between border control and rescue operations.[6] In what follows, the creation of this border bricolage on the Aegean Islands will be described, focusing on the blending of various surveillance practices.

The Emergence of EUROSUR

At first sight, EUROSUR and the hotspot approach may seem relatively unrelated. Although they were part of the same policy packages and were jointly applied during the migrant crisis, EUROSUR is mainly concerned with surveillance at sea, whereas hotspots are concerned with the identification and registration of migrants on land. However, the two programs are interconnected and share some specific characteristics. Both contribute to the intermingling of monitoring on land and at sea and the infrastructural integration of border security and human security approaches.

The development of the European Border Surveillance System (EUROSUR) is not only an example of an infrastructural compromise, but an example of infrastructural imagination as well. The Mediterranean is one of the main operational areas of EUROSUR. The system's main aims are to achieve interoperability between Europe's surveillance systems and increase situational awareness of critical events, predominantly border crossings. Surveillance at sea is central to border control and migration management in those parts of the Mediterranean that function as gateways between the European mainland, its southern islands, North Africa, and Turkey. Surveillance systems around the Aegean were already moving from a "patrolling driven" to an "intelligence driven" strategy before EUROSUR became operational.[7] The deployment of EUROSUR was supposed to Europeanize surveillance by extending the chains of associations between patrol boats, regional authorities, headquarters, and Frontex officials. Indeed, EUROSUR was introduced as a showcase of infrastructural technopolitics.

As a November 29, 2013, European Commission memo explains, the regulation establishing EUROSUR "provides a common mechanism for near-real time information exchange and interagency cooperation in the field of border surveillance . . . EUROSUR follows an intelligence-driven approach, allowing national and EU agencies to better understand what is happening at the external borders and to respond faster to new routes and methods used by, for instance, drug smuggling criminal networks."[8] The expansion of surveillance at the European Union's external borders affects its member-states in a number of ways. European framework programs and policy directives are implemented in national legislation and policy, accompanied by financial resources, technical support, staff, training programs, and technological and logistical infrastructures. But as technological systems are also locally embedded, coordinating existing border infrastructures can give rise to tensions.

EUROSUR is a computerized network for collecting, exchanging, and analyzing information for the surveillance of land and sea borders.[9] It can be seen as a "scopic mode of coordination" that accompanies the "mediatization of face-to-face situations."[10] EUROSUR fits the description of "infrastructural Europeanism."[11] Inspired by the notion of "infrastructural globalism,"[12] infrastructural Europeanism posits, among other things, that "the study of 'connections and circulations that have ceaselessly made and unmade different Europes' provides an entry-point into a new history of Europe that retells it as a genuinely transnational history instead of a collection of national histories."[13]

The transnational history recounted in this chapter is fueled by crises and conflicts and transgresses inter-European transnationalism. The EUROSUR program connects the European headquarters of Frontex in Warsaw with national coordination centers and regional authorities and border guards (in this case, the coast guards based in the Aegean Islands). Information and communication networks, based increasingly on the visualization of events, combine satellite imagining with radar detection and local observation to arrive at a picture of a specific situation, such as an unregistered ship entering national waters. EUROSUR thus brings together surveillance on land, at sea, and in the air.[14]

In 2015, we attended the European Day for Border Guards in Warsaw. When we asked a researcher working for the European Border and Coast Guard Agency (EBCG; then known as Frontex) about EUROSUR, he replied that there was too much buzz about it being a kind of all-seeing apparatus. He described EUROSUR more modestly, as consisting of "coordination, training, and funding."[15] Only later did we realize the import of his answer: monitoring mobility goes beyond technology, and technology goes beyond instruments. It requires protocols and personnel to gather, interpret, compare, and apply the information that is uncovered. The key word is "interoperability." But it also requires partnerships and collaboration among various authorities and institutes in many member-states.

This is where diplomacy and funding enter the picture. EUROSUR exemplifies how border infrastructures couple various systems of monitoring and control. From the outset, it was meant to be a combination of systems, a "system of systems," not unlike the Internet. The October 22, 2013, Regulation (EU) No 1052/2013 of the European Parliament and of the Council that established EUROSUR describes its aims as follows: "In the context of border surveillance, the ability to monitor, detect, identify, track, and understand irregular cross-border activities in order to find reasoned grounds for reaction measures on the basis of combining new information with existing knowledge, and to be better able to reduce loss of lives of migrants at, along or in the proximity of, the external borders."[16]

Originally a military term, situational awareness concerns "the ability to maintain a constant, clear mental picture of relevant information and the tactical situation including friendly and threat situations as well as terrain."[17] In border surveillance, situational awareness aims to visualize critical situations such as emergencies and irregular border crossings to assess whether intervention is required. In addition to boats, cameras, and radar, EUROSUR has been able to use satellite imagery, obtained through the European Satellite Centre, since 2014. Situational awareness is also an explicit goal of surveillance programs elsewhere, such as the US Secure Border Initiative Network (SBInet), that started in 2006 and was abandoned in 2011, which sought to install a state-of-the-art virtual fence along the US-Mexico border.[18]

EUROSUR is not only a system aimed to stimulate situational awareness in order to achieve real-time interventions. It is also a system designed to gather data about border crossings by migrants to fabricate maps of future migration risk scenarios. In that sense, EUROSUR contributes to the creation of humanitarian visibility. The notion of humanitarian visibility refers to "the regime of visibility shaped by humanitarian actors as one of the pillars of their intervention: that is, the thresholds and the mechanisms defining what must be seen and what can pass undetected or unnoticed."[19]

EUROSUR has a complex genealogy. Growing cooperation in border control and surveillance between national governments and EU institutions led to a greater emphasis on technology and the coupling of extant instruments and infrastructures. This development was also fueled by the rise of the European "homeland security market" and the security industry's initiatives to give shape to the technological border. A turning point here was the border package "Providing Europe with the Tools to Bring its Border Management into the 21st Century," presented on March 12, 2008, by Franco Frattini, the European commissioner responsible for justice, freedom, and security. On December 18, 2008, the European Parliament stated that it "welcomes the current discussions in the Council aimed at setting up the Eurosur border surveillance system with a view to optimizing the exploitation of all surveillance systems, essentially by extending their existing cover, which currently reaches only part of the areas where operations need to be carried out."[20] The emergence of EUROSUR can also be seen as the result of various small steps, of learning by doing and of copying successful practices from elsewhere. The emergence of EUROSUR is for instance linked to Spain's experience with SIVE, an "integrated system for external surveillance" that the Spanish government approved in 1999 to control the Strait of Gibraltar.[21] Seen in this way, EUROSUR is not a radical rupture in surveillance history, as it builds on existing technologies and networks. The development of this system is not the result of radical technological breakthroughs; rather, it followed an incremental, controversy-intense process. This does not mean that EUROSUR is "simply a reshuffling of preexisting elements"[22] Rather, it adds another layer to existing information and monitoring systems, knowledge practices, and communication networks.[23]

The Border Surveillance Laboratory in Action

As technopolitical projects, EUROSUR and the hotspot approach share a relationship with monitoring and visualizations. EUROSUR and the hotspot approach both perform monitoring, either by visualizing events (EUROSUR) or by identifying and registering migrants (the hotspot approach). There is even a deeper similarity here. Monitoring and intervening as well as vision and action are intimately related in border surveillance. Seeing is not an isolated activity of the eyes; it is bound up with locomotion with the movement of bodies and vehicles that carry the eyes.[24] Vision is embedded in the environment in which visualization operates—an extended and distributed infrastructure. While it is a precondition for action, vision requires previous actions and imaginations as well.

One way to emphasize this point is to underline the material basis of visualization, which often remains invisible in itself; we need to "specify the infrastructural work done prior to the possibility of rendering migration visible."[25] EUROSUR conducts a kind of "life governance" and enacts exactly this kind of "participation."[26] By combining interoperability with situational awareness, EUROSUR aims to connect all kinds of relevant information to undertake risk assessments of critical situations and to intervene in the "here and now." In that sense, the relationship between visualization and actual interventions suggests a comparison between the infrastructural setup of EUROSUR and that of a laboratory.[27]

As explained in chapter 2, the notion of a laboratory, particularly in science and technology studies, is much more specific than a test lab. The concept of a laboratory posits that to create a vibrant future for an innovation, the requirements include not only activity in the laboratory, but also activity outside the walls of the laboratory to prepare the outside world so as to shape a social order that will adapt this innovation.[28] As in a laboratory, surveillance systems aim to realize specific interventions by creating large-scale networks in which operations can be conducted to identify critical events. Laboratory practices tend to shift from representations to interventions, a movement that can also be seen in the EUROSUR laboratory.[29] But laboratories in this case are not fixed institutions; they are "centers of calculation" that gather data and information, relying on numerous mobilizations and enrollments to relate their findings to the outside world.[30] For border infrastructures more generally and for EUROSUR in particular, their networking capacity depends on myriad operations and mobilizations. As moving vehicles, borders not only consist of technologies and infrastructures, but they also operate as representations of critical situations and specific interventions.

Creating operability to arrive at situational awareness fits well within the program of virtualizing the border and visualizing critical situations. Nevertheless, at the end of the day, it is members of the coast guard who are sent out on patrol, to board boats and check unregistered vessels. In an interview, the director of Greece's Sea Border Protection Department in the port of Piraeus explained that border surveillance is only one step in the management of mobility.[31] This vision echoes the view that controlling the "means of movement" is part and parcel of modern state legitimacy.[32] There are several ways to actually enforce this control. In its most sophisticated form, surveillance comes close to "the politics of possibility," as Louise Amoore puts it. Central to this notion is the idea that the politics of surveillance is moving from a prevention-based paradigm toward a preemption-based one. The "anticipatory logic" of preemption does not seek "to forestall the future via calculation but to incorporate the very unknowability and profound uncertainty of the future into imminent decision." This logic informs what it means to draw boundaries—and to protect them. The sovereign capacity to write the borderline follows this logic by gathering "multiple elements of what is thought to be known about a person—each element, in the singular, a mere possibility . . . in order to give appearance to an emergent subject."[33] Both the border and the subject that aims to cross it take on the shape of a mosaic—the outcome of a process of "piecing together." The resulting "composite image" of the subject and the border line is produced as a kind of "mosaic."[34]

But the director of the Greek Sea Border Protection Department had something more mundane in mind. His proclaimed focus on the management of mobility and the movements of people tended to exclude the reasons why people migrate. The reality of surveillance then runs the risk of becoming a *hic et nunc* (here and now) reality, based on the monitoring of real-time movements to conduct immediate interventions. Under such circumstances, EUROSUR is not only far from a seamless web, it also is replete with tensions. In our interview, the union representative of the Hellenic Coast Guard at Chios sketched a depressing picture, not only of migrants but of coast guard officers as well. Border guards suffer high workloads and stress caused by shortage of personnel, long hours, a heavy emotional burden coupled with lack of psychological support, mounting paperwork, and lack of attention from superiors and authorities in Athens.[35] When the influx of migrants skyrocketed in 2014, coast guards were not prepared for their new tasks. Take, for instance, the simple, yet crucial—and in the end, complex—operation of stopping a boat that turns out to have undocumented migrants on board. The circumstances surrounding such situations were described by a Lesbos coast guard commander as follows:

> The Coast Guard's obligation is to protect and guard the sea borders, as well as the safeguarding of human life, and the respect of human rights. It can use all the available floating and land means, as well as aerial ones. It works in cooperation with Frontex, [nongovernmental organizations (NGOs),] and other international institutions. The Hellenic Coast Guard (HCG) is solely responsible for the coordination . . . The majority of the incidents in which we intervene has to do with

rescuing people. In this framework we have to cooperate with the Turkish Coast Guard. The goal is to stop them within Turkish borders . . . HCG's task is also to arrest the smugglers, who are usually on bigger, wooden boats . . . We use modern surveillance means in order to be able to locate such small boats during the night as well . . . In most of the cases the boats are located within Turkish borders and we inform the Turkish Coast Guard (TCG), which observe apathetically. In some cases, TCG even accompanies the refugees' boats.[36]

Proximity to the Turkish mainland, the bordering of Greek and Turkish waters, the "Europeanization" of border surveillance, and the volume of crossings by migrants have complicated the work of the Hellenic Coast Guard (HCG). Migrants, too, pursue strategic behavior, provoking rescue missions so as to be escorted to the Greek islands. Migrants frequently operate as "recalcitrant objects" by turning the means of surveillance back against efforts to "control the border."[37] Because perception requires action, migrants can make themselves visible in order to be subjected to action. When a migrant boat is observed, for instance, people on this boat can become the subjects of some kind of emergency. As such, "emergencies could be provoked by slashing a rubber boat with knives, jumping in the water, [or] capsizing a vessel. In any event, no longer does the patrol boat survey 'illegal migrants' and 'control a border'; it now witnesses an emergency and, as such, is legally obligated according to the same agreements that recognize sovereign territoriality at sea, to initiate a [search-and-rescue] operation."[38]

So how does one actually prevent a boat from entering national waters if it refuses to stop? The union representative provided us with some answers: shining a light, screaming, shooting in the air, or shooting at the boat's engine. But in the end, she told us, it all depends on the attitude of those in the boat. If the boat does not change course, there is hardly any way to make it do so. While EUROSUR may have created a border laboratory, this member of the coast guard felt like the mouse in the laboratory. "But I am not a mouse," she proclaimed.[39]

Hotspotting Greece

The combination of migration management, human security, and border security paves the way for several co-constitutions, such as between surveillance on land and at sea, and between monitoring and intervening. The border surveillance laboratory, a kind of "track and archive gaze," interacts and transforms different actors, institutions, technologies, and geographies.[40] The human security/border security nexus also stimulates the intermingling of care and control, a topic to which the next chapter will return in more detail. As argued in the opening of this chapter, the 2015 European Agenda on Migration created a particular infrastructural maneuvering space, which resulted from a compromise between the infrastructures concerned with human security and border security and led to a kind of humanitarian policing task of intercepting and rescuing migrants at sea.[41]

Although EUROSUR and the hotspot approach have different origins, they were jointly deployed by a specific peramorphic interplay. Interoperability and situational awareness are explicit goals of EUROSUR, but they apply to the hotspot approach as well. Whereas the EUROSUR program was to a great extent deployed at sea to combine border surveillance with search-and-rescue operations, the hotspot approach took place on land. However, both programs are connected. Although the origins of both approaches go further back, the deployment of EUROSUR and the hotspot approach was fueled by the conjoint 2015 European agendas on security and migration, in which a specific form of border management was announced to tackle the migrant crisis. The agendas aimed to install a compromise, a combination of monitoring borders and protecting migrants' lives, by connecting policies. The rationalities of the Schengen and Dublin systems encountered each other in a particular form of border management that combined surveillance and search and rescue with the registration of migrants at hotspots to process their asylum requests. The result, the following discussion will argue, was not only a multiplied form of border control, but also a hybrid form that blends the surveillance of landscapes and seascapes.

On October 15 and 16, 2015, Jean Asselborn, the minister of foreign affairs for Luxembourg (the state holding the EU presidency at the time), traveled to Lesbos with Dimitris Avramopoulos, the European commissioner in charge of migration and home affairs. The purpose of their visit was to see "how the hotspot is working as a pilot project in the process of being launched." During the visit, Avramopoulos referred to the efforts made by his country, while conceding that "there are significant shortcomings in the infrastructures."[42] Greece has been called the "gateway to Europe" because the conflicts in the Middle East have spurred an influx of migrants, most notably from Syria.[43] The number of migrants seeking to reach Greece by sea has swelled since 2014. The installment of the Evros fence on the Turkish border in 2012 did not stop the flow of migration, but simply redirected it. One of the most traveled routes by migrants from Syria and Afghanistan, as well as from African countries such as Eritrea and Somalia, entails trying to reach one of the larger cities on Turkey's western coast, such as Bodrum or Izmir. From there, they cross the Aegean Sea to one of the Greek Aegean Islands.

Greek migration policy in general, and border surveillance in particular, are often said to be ad hoc. The lack of resources, the difficulties in implementing policy agendas and legislation, and the murky interaction between central and regional authorities (e.g., between the capital Athens and the Aegean islands)—all lead to a provisional policy practice. The sustained economic crisis that has befallen Greece depleted the budgets of the ministries involved in border protection and migration policy. The already-limited resources of the police and coastal patrol officers involved in border control now faced a strict, internationally prescribed fiscal regime. The Greek state not only had to refinance itself and generate economic growth, but it also had to reinvent itself to modernize the governmental apparatus and restore its own legitimacy.

But it is not only the lack of financial and material resources that hampers Greek operational capabilities. Until recently, Greece did not have a detailed asylum policy; restrictive migration legislation allowed only a small number of migrants and refugees each year. Especially lacking were institutional (i.e., legal, social, and humanitarian) frameworks in which migrants saved from sea or detained on land could be cared for. Even before the euro crisis, it was questionable whether the Greek government would be able to receive such a large number of migrants over such a relatively short period of time. The combination of the country's economic decline, its highly porous borders, growing xenophobia, and faltering legal and institutional framework for the integration of migrants have created a fragile environment for the management of immigration.[44] Despite (or perhaps because of) the lack of institutional, professional, and financial means, the European Union has plowed resources into border surveillance in Greece. Especially since the building of the Evros fence, attention has been focused on the Aegean Islands adjacent to Turkey, which increasingly receive migrants from Syria and Afghanistan, as well as a flood of other migrants, mainly from Eritrea and Somalia.

The European Commission and its related intelligence and border control agencies have proclaimed Greek seas and islands, particularly those near the Turkish mainland, as hotspots. The hotspot approach has a different genealogy than EUROSUR, but it is programmatically closely related: the European Commission memo of November 29, 2013, that announced EUROSUR's launch also mentioned the option of installing "hotspots." The Commission describes the hotspot approach as follows: "A 'Hotspot' is characterized by specific and disproportionate migratory pressure, consisting of mixed migratory flows, which are largely linked to the smuggling of migrants, and where the Member State concerned might request support and assistance to better cope with the migratory pressure."[45]

The EUROSUR program and the hotspot approach share particular peramorphic mediations. Not only do they affect specific places on land and at sea, they create particular spaces of monitoring and governing. By gathering and analyzing data about migrant's movements and irregular border crossings, they not only execute border management, but also craft future-oriented risk scenarios. As Martina Tazzioli says:

> Hotspots pertain to "risk" levels as they represent the critical sites along the EU borders that are characterized by migratory events. Therefore, the visualization of migration in terms of level of risk contributes to the positing of a nexus between border (sites) and crisis: the term "hotspot" is used to designate critical spaces where there is need to intervene promptly to address a migration crisis.[46]

Just like EUROSUR, the hotspot approach is an infrastructural approach that connects the visualization and creation of a governance of risks with particular places, spots—hotspots. Whereas many analyses of the hotspot approach focus on the concentration and biopolitical governance of people in detention centers, the infrastructural angle emphasizes that hotspots are intimately connected with other programs of Europe's border politics.

As argued in chapter 4, detention centers can be understood as places where the state of exception is at work and where the sovereign power of states is enacted by biopolitically governing the bodies of migrants. However, the hotspot approach also combines a policy of containment and channeling mobility with the construction of data-driven control assemblages. This approach is conducted via a particular form of logistics, consisting of various steps of processing migrants and modes of infrastructuring. This processing and infrastructuring result in a "chain" of identification and registration that "moves not only migrants through containers, produces identities and data, [and] sorts files, cases, and fates into distinct institutional channels but also manages to coordinate different staff of national, European, and nongovernmental agencies."[47] Infrastructural differentiation and variation also create novel connections between mobility and containment and between surveillance on land and at sea. An example is the use of local health databases at ports to examine migrants immediately after they were picked up at sea.[48] But the hotspot approach not only concerns migrants, it also functions as a mixing bowl to stimulate cooperation among EU agencies. The hotspot is a mechanism that combines various European agencies to bolster their cooperation and centralize control over the common external border. The hotspot approach aims to unite various EU agencies and policies.[49]

The hotspot policies of the European Union have generated confusion. Are hotspots limited to specific places, such as reception and registration centers where migrants are fingerprinted following the Dublin Regulation? Or does the term more broadly encompass areas experiencing disproportionate migratory pressure? If the latter, the Aegean Islands could be considered a hotspot. Asselborn, Luxembourg's minister of foreign affairs, envisioned four specific tasks for hotspots: screening (to identify an applicant's nationality), debriefing (to analyze the applicant's travel route and identify smuggling networks), digital fingerprinting, and the provision of temporary authorization documents by the Greek authorities. The latter allows Syrian nationals to remain in Greece for six months; it is written in Greek and valid only for Greece. Non-Syrian applicants and those of uncertain nationality receive a document that allows them to leave Greek territory legally within 30 days.[50]

How do these procedures work out in practice? In an interview with the general director of the police in Chios, it became clear that there was a lack of, roughly, everything. Reflecting on the events of 2015, he recalled a lack of infrastructure, staff, and instruments, the most notable of which were Eurodac machines to take fingerprints.[51] Interpreters, staff to support fingerprinting processes, and members of Frontex and Rapid Border Intervention Teams (RABITS) only arrived in 2016.[52] Frontex teams usually consist of members of different nationalities. Apart from frictions caused by cultural differences (such as team members speaking their own national languages to each other), the police director thought that cooperation was relatively good. But then the role of Greece and the Aegean Islands emerged: "On 10 November 2015, an EU evaluation committee was on the island. They were here, in my office; it was like the Holy Inquisition. It was harder than passing the police officer exams. I felt as if we were a country that was in the evaluation process to become a member of the EU."

The mayor of the municipality of Chios emphasized not only the lack of people and equipment, but also the lack of a plan to anticipate future events:

> Understaffing and underfunding was the framework in which we had to act. The main problems the municipality encountered were the lack of a plan, which should foresee what we should do and the lack of infrastructures. The latter was not something irrational, as we didn't have the logistics to build infrastructures for receiving one thousand people per day. There was also the weakness of anybody to predict till when and how, as well as to give the specifications of what should be done.[53]

The influx of migrants dealt a hammer blow to the already precarious state apparatus. The mayor of Chios recounted:

> A major change that I've noticed is the change of the tasks of the police. Police have stop to care about anything else than the refugee issue on Chios. And this fact has major negative effects on social life. There are not even traffic police. And of course, there is no control on criminality, which fortunately is very low on Chios. This change of the state system has social consequences.[54]

The variety of tasks carried out by the local police and coast guard became apparent in an interview with the deputy head of the Chios coast guard:[55]

> We needed more personnel, for the registration, for locating the refugees on the coasts, for transferring them . . . We had to patrol a very big area. We had to transfer them for registration from very distant areas with the two HCG minibuses. . . . At the beginning of July and as the flows increased, we asked Chios Ktel S.A. to help us with the transportation from the beaches, as the police had already done.

Whereas cooperation with Frontex was considered satisfactory, the mayors of Chios and Lesbos were critical of EU highhandedness. According to the mayor of Chios, "each city has found a way to deal with the refugee issue, but the EU and national policies don't see this . . . I don't think that local societies can prescribe EU or national policy . . . in order to implement EU policy, a better connection has to be built up with the local societies."[56] On the subject of Lesbos, the mayor was more cynical: "We are worried about the way the EU is addressing the issue, which seems that it doesn't want to solve the problem, unfortunately. It wants to provoke problems with Greece."[57]

The picture is one of crisis management under precarious conditions, of responding to emerging events that nobody anticipated. Nor did anyone know for how long the situation would last. As the mayor of the municipality of Lesbos told us:

> The essential thing is that there is willingness to cooperate in order to manage the situation . . . It should be recognized that this crisis provides opportunities as well. And this opportunity is to restore Greece's dignity. We're not as they [the European Union] presented us, the bad Greeks, and they imposed us the Memorandums; we're people with capacities, maturity, dignity, and we have proven this . . . I think that the way I managed the situation on the island consists of a pattern that other mayors and the central government should adopt. For dealing with the emergency situation, I acted at the limit of legality.[58]

The installment of specific identification and registration centers based on the hotspot approach and the investments in surveillance at sea and support from Frontex and North Atlantic Treaty Organization (NATO) vessels in the waters between the Aegean Islands and Turkey were meant to strengthen interoperability and situational awareness during the migrant crisis. These programs combined monitoring and intervening, vision and action, by combining a focus on events at particular places with the creation of informational spaces. However, it remains questionable whether the EUROSUR and hotspot programs have succeeded in connecting the past, present, and future in order to anticipate migration flows. What they did manage to do was merge the surveillance of land and sea in novel ways and create new hybrid forms that fit into the compromise among migration, security, and border control.

The Merging of Land and Sea

Policies deployed by the European Union and its member-states have turned the Mediterranean into a sea under surveillance and the Aegean Islands into hotspots. They have done so by peramorphically creating a kind of border bricolage, in which improvisations, provisional arrangements, and ad hoc decisions are as important as the implementation and extension of European programs. The precarious compromises easily become compromised themselves.[59] The final sections of this chapter describe how this mingling of practices not only affects the practice of surveillance, but also how landscapes and seascapes come to be perceived and represented as ready for intervention.

The sea has become involved in the border politics of Europe as a natural entity that can either provide passage for migrants or be brought under control. "The border as vehicle" concept is shaped by the conflicting strategies of various actors and their attempts to visualize movements at sea, transforming how the sea is represented in maps, media, modeling, and migration management. The sea as a natural entity is increasingly divided up and absorbed in different forms of representation and control. Far from providing a seamless web, the monitoring of the Mediterranean manifests a seascape full of anxiety and risk.

The case of EUROSUR speaks to the question of how natural entities become visualized and represented in order to conduct interventions, whether they are humanitarian, aimed at border control, or both. By focusing on the sea as an area under surveillance, border infrastructures and monitoring activities spread to both national and international waters. In doing so, they confront a geographical, historical, cultural, and economic entity—the sea—with a variety of identities that preceded it becoming an area of security emergencies and humanitarian crises. Monitoring and surveillance at sea imply not only drawing pictures of risky and dangerous situations for humans, but also mapping the sea itself.

Unpacking EUROSUR through the study of its institutions and technologies reveals specific aspects of Europe's peramorphic mediations: the tensions among European, national, and regional levels of authority and coordination, as well as conflicts over defining geographical space. The transformation of the Aegean into an area under surveillance cannot be uncoupled from its history as part of the Mediterranean, a transnational zone of economic, cultural, and political exchange in terms of both cooperation and conflict. Studying the transformation of the Aegean under the aegis of EUROSUR highlights the roles that the sea, the islands, and the European mainland play in the politics of mobility, as well as how the actions of migrants, NGOs, international organizations, smugglers, and states transform both the landscape and the seascape into a particular domain of technopolitics.

The aim of creating a surveillance unity on the Aegean faces resistance from the region's multiple histories. While the central authorities in Athens are redefining the Aegean as a border region that functions as a line of defense against irregular migration, the sea and the islands are part of a long tradition of trade and cultural exchange and have more in common with spaces of interaction than places on the periphery now cast as the boundaries of Europe. The Mediterranean has a long history of dividing and connecting people, places, and cultures, but this identity is now interwoven with security policies. According to various historians, today's monitoring of the Mediterranean to manage human mobility and control migration movement fits into a long, verticalist tradition in which "the North" places itself above "the South."[60] Fernand Braudel's famous suggestion, supported by the French cartographer Jacques Bertin, to picture the Mediterranean upside down proposes seeing the historical sea as a collection of cultural, economic, and social trajectories and exchanges, a hybrid unity with a multiple history.[61] Over the past centuries, however, the Mediterranean has increasingly come to be seen as a European lake—a particular imagined geography that no longer functions as a middle passage or bridge from the South to the North, but is instead conceived of as a boundary.[62] With dramatic flair, Luigi Cazzato argues that the Mare Nostrum has turned into a Mare Monstrum: the "current militarization of the Mediterranean, which is trying to prevent the horizontal movement of migrants, is precisely a cruel (and probably pointless) attempt to impose 'unity' again, transforming 'bridges' into 'gates' controlled by one side only."[63]

Much of the scholarship on maritime governance and security has followed Braudel, who described the Mediterranean in terms of "the movements of men, the relationships they imply, and the routes they follow," conceiving of the sea as a plane of mobility and risks.[64] But sea should not be understood as the opposite of land, allowing exceptional representations and interventions. Although there is no such thing as *terra continens*—continuous or uninterrupted land—a philosophical understanding of terrestrial globalization is possible only if it is permeated by maritime stories.

One of the threads running through the work of Latour and Sloterdijk is the emphasis on the nonhuman and the posthuman. The networks and spheres they describe not only underline the material aspects of social and political order, but also function as containers of human thought and action. The delegations among social actors, institutions, and technologies contribute to changes in material morphology, as well as in moral and epistemological frameworks. The prefix "post" in "posthuman" does not refer to a period in time like "post–World War II" or a rupture in history like "the coming of postindustrial society." Instead, "post" indicates that the noun that follows it has somehow lost its meaning. Posthumanism does not mean that the end of humanity is imminent; humans will presumably still be here when robots perform our labor and algorithms make our decisions. Nevertheless, we are not uniquely important, and the building blocks of humanism such as autonomy, agency, responsibility, and human dignity may soon lose their meaning as moral and epistemological frameworks.

A similar development may be taking place in the context of border control. In the study of borders, this theme has been addressed in studies of border control and terrains at the US-Mexico border, to cite just one example.[65] An interesting example of how a border control landscape can be approached from such a perspective is the study of US Department of Homeland Security surveillance programs in the US-Mexico borderlands, which emphasizes "the quotidian role of a dynamic more-than-human landscape" in frustrating the department's enforcement practices and ambitions.[66] By unpacking the everyday challenges confronted by Homeland Security personnel, such research contributes to a posthumanist theory of terrain, shifting the focus of geographic inquiry to how the qualities of particular spaces, objects, and conditions may resist or impede routine navigation, centralized vision, and administrative practice. Interestingly, the notions of posthumanism and terrain not only sensitize us to the various relations and delegations between humans and nonhumans, but also emphasize how humans and nonhumans often fail to connect due to the resistance of the material or the frictions between them. A posthumanist theory of terrain "attends to the complex, textured dimensions of terrestrial space, and offers withdrawal—rather than association—as an analytic and ontological principle for theorizing a more-than-human political geography."[67] This argument is in line with "a post-humanitarian politics in which people, places, and things are engaged in contestations over mobility."[68] Posthumanism can be understood in these accounts as a notion that stresses the shortcomings of analyses that focus solely on humans, and of moral, ontological, and epistemological frameworks that avoid addressing technological, material, and geographical entities such as landscapes and seascapes in the study of border control. This posthuman perspective on terrain applies not only to landscapes, but to seascapes as well.

Borders are not just instruments that reach across various geographies; they also intertwine with them in spatial and material ways. This intertwinement can be conceptualized by conceiving borders on land, at sea, and in the air as solid, liquid, or gaseous, and distinguishing among three fields of action in border control practices.[69] The first, *solid* form of control is the border as a physical barrier, most often on land. The second, *liquid* form concerns border checks and practices of "policing and surveillance" involving processes of identifying, authenticating and filtering. The third, *gaseous* form is "the universe of the transnational database," connected to "the digital and the virtual, to data doubles and their cohorts, to categorizations resulting from algorithms, to anticipations of unknown behaviors, to the prevention of future actions."[70] EUROSUR introduces experiences from elsewhere into its fields of action, transforming control and surveillance by circulating methods, techniques, knowledge, and practices.[71] Surveillance on the Mediterranean "confers a new meaning upon Fernand Braudel's metaphor of the Mediterranean as an 'electro-magnetic field' in terms of its relation to the wider world":[72]

> The coasts of the Mediterranean, as well as state-operated vessels, are equipped with radars that scan the horizon around them by sending out high-frequency radio waves that are bounced back to the source whenever they encounter an object, indicating these "returns" as an illuminated point on a monitor.... Optical satellites generate imagery by capturing reflecting energy of different frequencies such as visible and infrared light, while satellites equipped with synthetic-aperture radar (SAR) emit a radio signal and create an image based on variations in the returns. Both "snap" the surface of the sea according to the trajectory of orbiting satellites and are used to detect unidentified vessels.[73]

The crossover between fields of action and control in different border areas merges the border landscape with the border seascape. Although different instruments and techniques are applied on land, the "liquefaction" of the surveillance field continues when migrants disembark from their boats and step onto land.[74] Conversely, the intensified surveillance at sea—particularly operations to prevent boats from entering and pushback operations—echo the solid approach generally associated with the control of land borders. Given the geography of the borders of Chios and Lesbos, nature enters the picture as a material entity.[75] The hybrid role of EUROSUR is not restricted to assemblages between humans and information networks; it also includes a mingling of technologies with different geographies.

The Border Bricolage

The migrant crisis of 2014–2016 revealed the intensity of surveillance projects both on land and at sea. The combination of security, migration, and humanitarian approaches of the European Union resulted in the rise of distinct but related projects, such as EUROSUR and the hotspot approach. Connecting surveillance to search and rescue also resulted in the blending of the geographies of land and sea in unforeseen ways. This chapter followed the peramorphic mediations between elements and situations so as to analyze the mutual transformation of technologies and territories.

As Latour does with his notions of "association" and "translation," this discussion has identified the points at which these transformations took place. While the EUROSUR program and the hotspot approach predate the so-called migrant crisis, both found timely application and a field of operations in the Aegean Sea and islands. But instead of providing a seamless overview, the surveillance was highly improvised and had to strike all kinds of compromises under often-trying circumstances. Things turned out differently than foreseen in many ways.

The composition of border infrastructures consisted of the intermingling of seeing and acting on land and at sea and the merging of the respective geographies. EUROSUR coupled all kinds of instruments from different member-states to create interoperability and achieve situational awareness, relying on all kinds of ad hoc arrangements and a specific interplay of seeing and acting, representing and intervening. To close the holes, places are required where what Latour calls a "panorama of associations" is created and local activities become a bigger issue.[76] Like the previously described centers of calculation in science, coordination centers in the field of border surveillance "act at a distance" by producing all kinds of forms and standards that allow connecting activities.[77] This mélange of agents, institutes, and technologies not only allows the multiplication of entities in data, information, facts, and representations, but also allows their black-boxing through standardization as a way to close disputes. The hotspot approach connected registration centers, regulations, and instruments with local procedures and provisions in what was commonly referred to as a "crisis." As a result, the border infrastructures at Lesbos and Chios in 2014–2016 saw a multiplication of not only actors and agencies, but also the geographies of land and sea.

Greek migration policies appeared to be chaotic, ad hoc, and mainly based on improvisation and crisis management. But much the same could be said for the European approach to its external borders in the same period. Even the most sophisticated, prudent, and technologically advanced approach would not have prevented the fragmentation of practices at the borders. Border infrastructures travel across landscapes and seascapes, relying on myriad technologies to address various kinds of border events. While the transformative strength of border infrastructures sometimes manages to bridge these differences, technologies do not merge smoothly, and gaps keep appearing, created by frictions that are inseparable from the aims of creating interoperability and situational awareness; by the geographical resistance of landscapes and seascapes; and by the migrants themselves, who refuse to be the passive objects of migration policy. A border infrastructure can be seen as a composed complex that comes into existence through various activities, hanging together and rendered actionable through a number of technopolitical compromises.

Compromises are the result of peramorphic mediations. These mediations do not just concern the interactions among actors, institutions, and technologies. Mediations also lead to articulations (or, as Sloterdijk would say, "explications") to the emergence of new configurations because they affect and infect each other. The notion of "peramorphic politics" that was introduced in the beginning of this book emphasizes the expansion and multiplication of borders. Both EUROSUR and the hotspot approach drew attention to particular events at specific places, but they also stimulated the emergence of informational spaces. EUROSUR and the hotspot approach contributed to the building of particular archives of information about the mobility of people, as well as to the constructing of windows of action that allow agencies to examine, identify, and register migrants. The ongoing process of creating borders affects everything it touches. The compromise between migration, security, and border control, as well as the aim to combine the protection of human lives with the protection of borders by putting surveillance on land and at sea together with search-and-rescue efforts and identification and registration, multiplied the variety of border entities. This multiplication of borders not only leads to an expansion of borders, but creates ever more hybrids: it merges land and sea, vision and action, mobility and containment, monitoring and intervening, and, as chapter 6 will show in more detail, care and control.

Belongings of migrants in a camp in Serbia, May 2015.
Source: Henk Wildschut.

6 The Portable Provision of Care and Control

The Mediating Humanitarian Border

Greenpeace and Médecins sans Frontières (MSF) running joint rescue operations; volunteers and humanitarian professionals working on land and water; workers with the United Nations High Commissioner for Refugees (UNHCR) cooperating with Greek police to run reception and detention centers; Greek coast guards and ships from European Union (EU) member-states joining a Frontex operation; cooperation between Frontex and NATO—the emergence of border infrastructures on the Aegean Islands during the so-called migrant crisis of 2014–2016 consisted not only of state initiatives to patrol the border, but the provision of food, shelter, medical care, and legal assistance. Local villagers and foreign volunteers offered acute medical care. Others entertained the children after their long journey. Still others helped migrants get to the buses that would transport them to the reception centers. Along with the support from these volunteers, organizations, and agencies, clothing, goods, and other necessaries, ranging from infant formula and sleeping bags to rucksacks and raincoats, were sent from Greece and all over the world to Lesbos and other places.[1]

Whereas chapter 5 analyzed the technopolitics of EUROSUR, the hotspot approach, and the way that various territories are blended by surveillance activities, this chapter turns to the fusion of providing care and enforcing border control and the compromises made between security and humanitarian care that underpin them. The particular manifestation of the border that this chapter will analyze is the humanitarian border as it materialized at the islands of Chios and Lesbos during the migrant crisis of 2014–2016.

A humanitarian border combines border control policies and the presence of state agents with the provision of medical expertise, medical care, and legal know-how (such as seeing to human rights issues and providing interpretation and translation services).[2] Humanitarian initiatives and initiatives to regain border control—often under the umbrella of a security agenda increasingly interwoven with migration policy—have a reciprocal relationship. To a certain extent, the combination of humanitarian support and security policies can be seen as a merging of national security and human security approaches.[3] A humanitarian border emerges out of particular peramorphic mediations that can be classified as "borderwork" or "humanitarian borderwork."[4]

Humanitarian borders existed before they were coined as such. Sangatte, for instance, a transit center run by the Red Cross in the north of France near the port of Calais, where many migrants on the way to the United Kingdom were stranded, could already be considered a humanitarian border. Calais is an important location for migrants who aim to reach the United Kingdom, by boat via Calais or by train and motortruck via the Chanel Tunnel. After the French minister of the interior (and future president) Nicolas Sarkozy closed the refugee camp in Sangatte in 2002, migrants started to build encampments in the woods, often referred to as the "Calais Jungle." The situation at Sangatte and in the camp at Calais was the scene of much tension between "securitization" and "humanization," between "repression" and "compassion."[5] The transit center itself reflected this tension: it was a place of "indeterminate status," as it was neither a proper reception center nor a detention camp. The center was staffed by the Red Cross and the French police.[6] Although the circumstances are different, there are many similarities with the hotspots in the Aegean Islands, and these will be discussed in this chapter, particularly the intermingling of care and control.

Humanitarian action by nongovernmental organizations (NGOs) is based on two interrelated meanings of the concept of "humanity." First, this concept suggests that humanity consists of an indivisible collective of humans. Second, it contains a certain willingness of people to show humanity to others who suffer. This unity between humans, consisting of recognition, compassion, and solidarity, is not free of friction or devoid of distinctions or inequalities. On the contrary: "as one gets deeper into humanitarianism a series of dimensions of what may be called a complex ontology of inequality unfolds that differentiates in a hierarchical manner the values of human lives."[7]

The tensions that vibrate within humanitarianism are foreshadowed by three characteristics of humanitarianism as it arose in the nineteenth century. Humanitarianism is distinct from previous forms of charity, compassion, and philanthropy in three respects. First, it aims to cross boundaries. Second, it has transcendental significance. And third, it implies a certain form of governance that connects compassion with control.[8] The humanitarian border at the Aegean Islands, which will be discussed in the following, particularly reflects the last point.

The connection between compassion and control that results in the emergence of humanitarian borders entails specific forms of peramorphic translation work. The humanitarian border shares some specific similarities with the border infrastructures that have been described in previous chapters of this book. These similarities concern the compromises that are required to connect various political ideas, actors, institutions, and technologies to make a border, and the mechanisms that underpin the creation of a technopolitical entity that is a border.

The particular peramorphic mediation that will be examined in this chapter is the relationship between humanitarian aid and security—between care and control. As the previous chapter explained, the emergence of EUROSUR at the Greek Aegean Islands during the so-called migrant crisis can be considered an example of an infrastructural compromise, a compromise between security and humanitarianism. The following discussion will explore the intermingling of care and control as an example of an infrastructural compromise. It focuses on two issues in particular. First, the humanitarian border will be conceived as a movable configuration that arrived with the migrants that went to the islands of Lesbos and Chios. Second, the analysis will show how this humanitarian border is only precariously attached to the Aegean Islands, requiring all kinds of exchanges with local authorities, inhabitants, officials, and volunteers.

The analysis provided in this chapter is partly based on a series of interviews. In order to gain insight into the entanglements between care and control, thirty interviews with various Greek authorities, representatives of Greek and international NGOs, grassroots organizations, and local volunteers were conducted on the islands of Chios and Lesbos between February and May 2016.[9] The interviewees were asked to reflect on the tensions and entanglements between border control (e.g., Hellenic Coast Guard patrols, Frontex operations, and the hotspotting policy to register migrants) and humanitarian aid and rescue (by NGOs, local grassroots organizations, and volunteers) at the height of the local migrant crisis between the spring of 2015 and early 2016. Excerpts of the interviews are employed in this chapter to illustrate the mobility of all kinds of goods, things, and devices, the interactions between the various organizations, and the tensions that arose as a result.

The Making of Movable Borders

One way to reconstruct the emergence of the humanitarian border is to follow the migrants and to analyze how, step by step, they became part of a configuration that can be called a humanitarian border. "Becoming part" does not mean that this border constellation is already a fixed entity. To a certain degree, migrants contribute to the shaping of a humanitarian border, as their movements already shape particular form. For instance, the volunteers, professionals, and coast guards who follow the routes of migrants can be said to form a kind of "corridor."[10] This is not the same as claiming that migrants carry the border with them. The statement that borders are somehow attached to migrants may prompt comparisons with the Titan Atlas carrying the world, drawn with borders, on his back—or worse, Sisyphus rolling an immense boulder up a hill over and over again, unable to escape this punishment. But unlike these mythological metaphors, borders and migrants are not inseparable; theirs is a dynamic relationship. Nor are migrants the sole actors involved in the traveling of borders. Solely focusing on migrants would be overstating the role of one kind of actor in a field crowded with organizations, legislations, institutions, and materialities. Nevertheless, placing migrants as the center of attention provides a counterweight to state-centric perspectives and may open up novel ways to analyze the border infrastructures they are part of. Then again, the study of *who* travels must be accompanied by the study of *what* travels. Migrants arrive by boat; and once on the islands, they are provided with food, medical care, and documents. They must be transported, sheltered, and taken into detention. The provision of care and control consists of the circulation of all sorts of things, technologies, and materialities.

Underlying this analysis is the idea that borders are not only created by the authority, sovereignty, and jurisdiction of states. First, Greece is part of the European Union, and many of its border policies are in fact EU policies—or even European policies, as argued in the opening chapter, because these policies are not restricted to EU member-states or members of the Schengen Area. Second, states are but one actor in a variegated landscape of international NGOs, volunteers, and UN agencies. Third, and most important for this chapter, borders arise out of various and often overlapping or even contradicting infrastructures. Care and control, fixation and motion, centralized and dispersed: all of these opposites apply.

The border infrastructures that arose on the islands of Lesbos and Chios can aptly be seen as a kind of "archipelago."[11] As a group of islands, the Aegean Islands indeed form a particular kind of border area. On the other hand, borders are not only about isolation, but also about connection. In addition, islands can point to flows of people, movements among islands, the relations between water and land and between islands and the mainland. The comparison of borders with islands can be understood more intimately if we emphasize the isolation and connections between islands, as Peter Sloterdijk did in chapter 3 when speaking about foam structures. Once again, the "border as island" requires us to consider the movements of both people and things; and again, applying Latour and Sloterdijk's perspective leads to a blurring of dichotomies. The emerging borders on the Aegean Islands neither isolate events nor encourage continuous flows; rather, they lead to the coming-into-being of all kinds of pop-up border infrastructures on location. Following the notion of immanence espoused by Latour and Sloterdijk in chapter 3, the border as island will not be studied from the outside in, but from the inside out.[12] The border as island and the island as border do not lie on the boundary between the inside and outside of Greece and Europe, or between the European mainland and the Mediterranean. Instead, the traveling border arrives on the island with the migrants and organizations and people engaged in the various provisions of care and control.

Care and Control at the Humanitarian Border of Lesbos

A rectilinear reading of the events on Chios and Lesbos, when the number of migrants seeking to reach Europe peaked in 2014–2016, would be that the Greek authorities, encouraged by EU policies and Frontex assistance, tried to get the situation under control, while international NGOs, refugee organizations, local residents, and volunteer groups tried to provide the migrants with humanitarian care. But in practice, things turned out to be less obvious and, most of all, less dichotomous. After the Greek elections of January 25, 2015, the Coalition of the Radical Left, Syriza, formed a government with the nationalist conservative Independent Greeks Party, known colloquially as ANEL. Although before the elections, Syriza championed the opening of borders, taking down the Evros River fence, ending the pushback operations, abolishing detention centers, reformulating national asylum policy, and instituting safe passage for refugees, the new government quickly changed track. In the meantime, the presence of national and international NGOs and volunteer groups was gradually growing on the Aegean Islands.

The combination of EU hotspot policies, migration management by the Greek authorities, and support provided by NGOs and volunteers can be regarded as an emergent humanitarian border. It must be stressed that the notion of the humanitarian border is not just an invention of scholars. The International Organization for Migration (IOM), the UN migration agency, states on its website:

> IOM observes a strong need to protect the human rights of migrant populations during any crisis, particularly in those that results in cross-border movements. Officials at the border are usually the first to be confronted with such unusual movement dynamics and border security can become affected. Well-prepared and managed crisis response at borders can improve humanitarian action and protect vulnerable migrants while maintaining the security of states and borders. Through its humanitarian border management approach, IOM assists governments and their border institutions to more effectively prepare for and manage crisis-induced displacement and mass movements.

An example of such a humanitarian border is a reception center, such as the ones on Chios and Lesbos. Not only is the term almost a combination of the functions it houses (namely, registration and detention), it also presents them in the name of hospitality. My first encounter with the humanitarian border in a reception center was in 2014.[13] The hotspot policies, as described in the previous chapter, had not yet been implemented, but with the assistance and support of EU and Greek funding, the Moria refugee camp on Lesbos had taken shape on a former military base. When we visited the place, only the small camp was open, although a larger camp that would open under the hotspot policies had already been built. The architecture and spatial organization of the camp reflected the division of roles of the people working there. One part of the camp housed the NGOs: Medicins du Monde (Doctors of the World), UNHCR and Meta, the translators. In another part of the camp, Greek police and Frontex officials kept an eye on the detained migrants, mainly Syrian refugees.

The peramorphic mediations that lead to the coming into being of a humanitarian border are not a step in a linear process, in which borders penetrate other domains of life and governance: as Walters (2011) says, "to focus only on new developments in surveillance and control risks a rather linear and developmentalist narrative." The notion of the humanitarian border "is not just to insist on the emergence of a domain which deserves to be taken seriously in its own right. It is also to complicate the linear narrative; to suggest that at the same time that borders seem to become more like this, they are also taking other forms, materializing along other lines whose trajectory is difficult to predict."[14]

Part of this materialization is due to "humanitarian borderwork" which "introduces explicitly humanitarian actors into the borderwork assemblage."[15] But borderwork consists of more than human actions. According to Jones et al. (2017), "As humanitarian borderwork introduces new actors, it also works to produce new types of border spaces constituted around practices of rescue and the provision of basic needs while introducing new categories of life and consolidating socio-political hierarchies."[16] This borderwork takes place not only in the registration and detention centers, but also outside these hotspots. Moreover, there is a clear chain of associations that connect the inside with the outside. See, for instance, this description of the situation outside the Moria camp in October 2015:

> Volunteer organizations set up an "exterior" camp in October, on a sloping olive grove immediately adjacent to Moria. Though the infrastructure of this overflow area is considerable—featuring a clinic, child-friendly area, enclosed restrooms, mosque, kitchens, and tea distribution center—problems are legion. Human Rights Watch Emergencies Director Peter Bouckaert, in a November visit, observed asylum

seekers sleeping outside, surrounded by squalor, crowds jostling and fighting for access to the registration center, and parents and pregnant mothers fearing for their safety and that of their children. For want of rigid-structure shelters, asylum seekers in the exterior camp are housed in tents of varying size and quality that provide limited protection from wind and rain and little insulation from winter temperatures. In late December 2015, Starfish Foundation began distributing firewood to give residents an alternative to burning garbage in order to keep warm.[17]

Hotspots are but one node in the network of dispersed humanitarian borders in the Aegean Islands. The humanitarian border—and its attendant humanitarian borderwork—are highly mobile; they do not create a fixed infrastructure. An example of this mobility is how migrants arriving on Lesbos might encounter MSF at different times and locations:

> In December 2015, migrants arriving in Lesvos would have first encountered MSF at sea as they engaged in what can best be described as pre-emptive SAR missions. Here, MSF along with partners from Greenpeace worked from a rigid-hull inflatable boat to monitor and guide the rubber dinghies and other small vessels making the 10 – 12km crossing from Turkey to safe places of disembarkation on the beaches. Migrants would then encounter MSF again when they took one of the MSF-contracted buses from the village of Skala Sikamineas on the north-east shore. After a 20 km drive, they could wait, sleep, get warm, pick up blankets, receive information or use the Wi-Fi network at the MSF-run transit point at Manatamados, from where they caught another bus to the Moria "hotspot" to register with the Greek police and Frontex. At Moria, people may have received basic medical triage from MSF medical practitioners working inside the hotspot.[18]

If it is true that borders materialize where the migrant is, it should be possible to distinguish particular entanglements by "following the migrant."[19] The daily practices of the migrants arriving on the Aegean Islands, the organizations and volunteers supporting them, and the various agents trying to manage migration are interwoven with things of all sorts. These include, for instance, the ships on which the migrants arrive; the cell phones that they use; the housing, food, clothes and medical care that they are provided; the fingerprinting machines used to register them; and the buses, camps, and centers deployed to detain them. At one point, migrants interact with the Hellenic Coast Guard that is charged with detecting smugglers; at another, they collect shoes, clothes, and baby things from Médecins du Monde. The humanitarian border consists of all kinds of materialities and is woven together by discursive as well as material engagements by a variety of actors.

A telling example of such intermingling is the site of the volunteer- and refugee-run PIKPA camp near Mytilene Airport on Lesbos. I visited PIKPA in September 2014.[20] While PIKPA is under the umbrella of the *Xorio tou oloi mazi* ("All Together Village"), in which several other local grassroots and NGOs participate, it has ties to the authorities as well. The Municipality of Lesvos pays the water and electricity bills, while some of its volunteers are registered so they can enter the detention center at Moria. PIKPA volunteers provide daily dinners in the Moria detention center, while refugees who stay in PIKPA cook their own food in a kitchen supplied by a donation from an organization. PIKPA volunteers are also active on the coasts. In an interview, a member of PIKPA told us how relations with the Greek authorities are dispersed around the island:

> The regional authority doesn't exist; they're completely absent. There were tensions in the south of the island where centers were planned, mainly in Molyvos . . . We proposed to build a registration center in the southern part of the island, as half of the refugees arrived in that area. They didn't want it. Community people's assemblies were held and Molyvos' residents rejected the proposal . . . Then the Petra[21] ex-military camp was proposed to be turned into a registration camp, located at the entrance of the village. They tried to transfer some small houses there, although the residents blocked the road. The residents were on shifts for three nights . . . We also proposed Molyvos municipal campground, which is close to Molyvos, and it's been closed for several years . . . At first, in May–June, they started to host refugees in the parking lot next to the school and the residents chased the refugees; they also closed the school for a day, arguing that their children were there and they'd get sick. Several refugees stayed there and volunteers tried to help despite the big pressure they had from the residents. Then the tourist season started. The residents rejected having a structure built there, saying that if you create the infrastructure, then it's as if you invite them.[22]

Meetings between voluntary organizations and state authorities often feature discussions about the moving of concrete things such as centers, camps, and campsites. This adds another dimension to peramorphic politics, as this kind of politics is apparently concerned not only with objects that move, but also with things and places that all of a sudden become involved in the policies of movement.

While care and control are infrastructurally interwoven, how can the resulting configurations be said to be movable? Objects are usually not granted any agency. Instead, they "are almost always 'used' or guided by human actors, assumed to possess full agency."[23] The commander of the Lesvian coast guard told us in an interview that "the expertise on registration, fingerprinting, and identification processes" (i.e., how he characterizes a hotspot) is helpfully concentrated where people arrive.[24] But the focus on *who* and *what* travel should not prevent us from considering the relationships with other agents, institutions, and technologies. "Following the migrant" as a strategy to understand the configuration of care and control would fail to capture the coordination between NGOs, grassroots organizations, the police, the coast guard, and local and national authorities. Humanitarian initiatives are thus best understood in relation to security measures as they materialize in mutual interaction. A strict focus on materialities would keep the practical engagements that actors have with these materialities out of sight.

Cooperation, Coordination, and Conflict on Chios

Focusing on the materialities and the movements that constitute a humanitarian border is not only a means to detect the shaping of infrastructural compromises; it also allows a view that center-stages the movements of migrants themselves. As the previous discussion clarified, this view does not claim that migrants should be the only actors that get followed, nor should they be regarded as the cause of the creation of a border. The idea that borders travel with migrants is not to be mistaken for the view that the border simply materializes where the migrant is.[25] To a certain extent, migrants and the borders they carry with them work as magnets.[26] A freelance journalist on Chios similarly told us that humanitarianism "creates niceness where refugees are."[27] But the point of the traveling borders perspective is to emphasize that actors engaging with migrants do not operate in a single overarching network, in which tensions between humanitarian actors and state officials are smoothed over by reciprocal adjustments of conduct. Instead, different networks exist simultaneously. They coexist.[28] Focusing solely on migrants is too narrow a perspective.

The networks of actors engaged in activities of care and control may seem to exist side by side. But they actually intertwine and overlap, with actors who can circulate in both networks, mediating between them in mutually coordinated conduct. The mayor of the municipality of Chios told us that most of the coordination is done voluntarily. This willingness is important to him, as it emphasizes the flexibility of many of the organizations and volunteers—a flexibility that state agencies lack. But this flexibility does have its limits; at a certain point, said the mayor, you have to "ensure stability."[29] But ensuring stability often leads to tensions with international NGOs, grassroots organizations, and local volunteers.

Voluntary actions begin as soon as migrants arrive. A volunteer from Agia Ermioni, a fishing port 10 kilometers southeast of Chios town where many refugee boats arrived, explained:

> The volunteers started welcoming the refugees, providing them with dry clothes and shoes, food, mainly biscuits and so on, tea and baby milk, for as long as they were waiting for the buses to transfer them to the registration center. The boats were arriving during the night as well, at 3:00 or 4:00 in the morning, and the volunteers were there helping them. The refugees were mainly families with children and babies. On some days, more than 300 refugees arrived. In each boat there were more than 60 people. Men were sitting on the sides of the boat, the women and children on the boat's floor. The majority of them were wet when they arrived, as the boats were overloaded. At the beginning they were using the local association's small house, but as the flows increased, they built a changing room and another small house, using the wood from the refugees' boats. They needed space to put the clothes in order and to dress the children, especially when it was raining.[30]

The mutual dependencies between volunteers, organizations, and state agencies were emphasized by a volunteer at Chios Solidarity, a group that arose out of the initiatives in the central gardens of Chios town: "We cannot work completely independently. We don't have close cooperation, but a basic communication and exchange of information on needs that we cannot cover which they could cover, and some others that they cannot cover and we provide."[31]

The configuration of the practices of humanitarianism and security materializes out of two-way traffic. A member of Lathra—a grassroots refugee solidarity committee on Chios launched in 2001—told us that their activities have taken shape in iteration with those of the coast guard.[32] In the years that Lathra has been present on Chios, there have been several disagreements. Lathra now refrains from actions that may create problems with the authorities and works with the coast guard to define possibilities to act.[33] The entanglement of the practices of humanitarianism and security also materializes in humanitarian actors performing acts of control and security actors performing acts of care. The area manager for Greece of the Norwegian Refugee Council,[34] an independent humanitarian organization that helps people forced to flee in various countries, stated: "The idea is to put a bit of oil in all the mechanisms. It's to push the different authorities to work together." He went on to explain:

> Sometimes the police is not working with the Greek Asylum Services; sometimes you have one decision taken at the national level which doesn't really fit with the local environment, although the objective . . . cannot be reached in the way the order has been given. So the authorities always have to find a way to be pragmatic in order to reach this objective, but maybe also to take into consideration the overall environment which includes different services, Greek public services, European services now that we have EASO, Frontex, etc., and of course the interests of Chios civil society and the interests of the migrants.[35]

The Norwegian Refugee Council provides many kinds of care—water, waste management, a shuttle bus, the distribution of clothes, and information about legal frameworks and the management of camps. Its representative told us that they try to "connect the dots between the authorities." For example, "in Souda and Dipethe [former refugee camps] we did vulnerability profiling using the registration list of the police."[36] A consultant for the Ministry of Migration commented on this practice: "In any case, the logic is that at the first stage we register, identify, detect the vulnerable groups and the asylum seekers, and at the same time we detect the people who should be sent back."[37] An actor whose primary responsibility is providing care thus simultaneously performs practices of control, in part confirming the observation that "discourses concerning the human rights of asylum seekers are de facto part of a securitization process if they play the game of differentiating between genuine asylum seekers and illegal migrants, helping the first by condemning the second and justifying border controls."[38] These shifts in roles and positions also occur at sea, such as by NGOs that rescue migrants from their boats:

> By cooperating with Frontex and Eunavfor Med, as well as by transferring people and handing them over to the Italian police authorities, NGOs are not only relieving governmental actors from their responsibilities. They are also providing operational support and humanitarian non-state legitimation to the border regime they declare to contest. Like those aid workers who "become logisticians in the war efforts of warlords," they thus become part of a hybrid border management system that results in limiting the freedom of asylum seekers in Europe through the Dublin Regulation, in either forcibly returning or illegalizing those who are denied protection, in arresting and prosecuting purported smugglers, as well as in enhancing cooperation with countries of departure with the aim to prevent people from reaching Europe.[39]

Security practices and effects often arise out of humanitarian governance. Many humanitarian actors perform acts of control. But security actors similarly perform acts of care. The manager from the Norwegian Refugee Council also noted that the coast guard and the police "push for solutions in the best interest of migrants . . . they have a positive approach . . . [although] they are not supposed to be that positive." When it comes to patrols at sea, he was "sure that they have saved a lot of lives, and I'm sure that they were doing their job properly. . . . My feeling is that this [i.e., doing pushbacks and treating migrants improperly] is not what they want to do. There is a good spirit at least here in Chios."[40] This was confirmed by the former deputy head of the Chian coast guard: "If you are in the borderline, the only thing you can do is rescue people."[41]

Nevertheless, conflicts seemed unavoidable, such as when the provision of care came to be seen as undermining solidarity on the island. This occurred when, according to a member of Lathra, "solidarity groups, the people who were helping, were suddenly presented as something dangerous, something which creates problems, something suspicious and selfish."[42] And there are also limits to what NGOs can and will do. While MSF's decision that it will no longer accept funds from the European Union and its member-states following the EU-Turkey Statement[43] has received the most attention, volunteers at the Chios Social Kitchen had already refused to distribute food in the Vial detention center.[44]

From this discussion, we can conclude that apprehending the coming-into-being of the border from the inside out requires a more variegated repertoire than Latour's associations and Sloterdijk's foamy configurations. The tensions, frictions, contradictions, and consequences of the various border events result in highly dispersed configurations that appear and disappear. The following engages with the question of how conflicts and controversies in these border infrastructures can be understood in greater detail.

The Becoming of a Border

Border infrastructures bring together a number of actors, including migrants, state agents, and people working with NGO programs to monitor, register, and surveil with humanitarian support. A symmetrical point of view entails unpacking the institutional structures, technological and material assemblages, and emerging migrant configurations simultaneously. This perspective would suggest that migrants, while not the only subjects on the move, provoke the coming-into-being of border infrastructures as all kinds of other actors, organizations, and institutions respond to their arrival. It would also question entrenched dichotomies, such as between nonstate actors providing humanitarian support (care) and state actors addressing migration as a security issue (control). Finally, symmetrical treatment may allow a more nuanced view on the scale and scope of the agentic capacities of actors and help unravel all kinds of tensions among the various actors, institutions, materialities, and technologies.

The emphasis on materiality is not meant to reduce border infrastructures to their material components, but rather to inform *becoming*. The emphasis on materiality adds specific content to the notion of *work* in humanitarian borderwork.[45] While materialities may accelerate or slow down these processes of becoming, they are inextricably linked to spatial and temporal processes of circulation. A symmetrical perspective on border infrastructures privileges neither stable border infrastructures nor migrants as actors. Instead, it seeks to unpack the movability of border infrastructures by revealing the interactions among institutions, technologies, and migrants. One consequence is that neither technology nor materiality is a sufficiently suitable starting point. Instead, border infrastructures must be studied at the moment of their making. Moreover, the emphasis on networked technologies or the state's border infrastructures tends to neglect the things that matter most to migrants, including the materials they are equipped with or confronted by in daily life.

Among the characteristics of the humanitarian border, one is of particular importance: "the humanitarian border is not a fixed border but something which fluctuates. Its geography is determined in part by the shifting routes of migrants themselves."[46] The close examination of various technologies and materialities indeed reveals that things move in different ways, and sometimes they do not move at all. For instance, a member of the Starfish Foundation on Lesbos emphasized the organization's local roots. The specificity of events on a *particular* stretch of coast near a *particular* village informed the materialization of the initiative and remained its focus over time: "We are not like other organizations who come here to help and when it gets difficult they will be like, 'OK, now we are going to help somewhere else.' This is our place."[47] The mayor of Chios similarly told us that "the local self-organized initiatives, mainly on the coasts, were set up because people arrived in front of their doors,"[48] again stressing the local context that directly shapes the materialization of the humanitarian border.

In contrast, Médecins du Monde is an independent humanitarian movement working to empower excluded people to access health care around the world. Through 400 programs in eighty countries run by more than 3,000 volunteers, it provides medical care, strengthens health systems, and addresses the underlying barriers to accessing health care.[49] Médecins du Monde has a mobile unit, which a staff member on Chios described as follows: "The mobile unit moves to places where there are needs of medical care . . . The goal is to cover more points . . . It's not necessary to be properly settled."[50]

An interviewee from Starfish was critical of NGOs that hop between locations. But although she claimed that the locality of its work makes the foundation more sensitive to its immediate context, its focus on the local at times proved to be a weakness: "At the end of August, the area of the bus stop was closed down because the school was going to open and people wanted the refugees out of the village. So, the refugees were arriving here on the coast, and this is the road you have to walk to Mytilene [the harbor and capital of Lesbos]. So, in the village here, the first village they would arrive at, they would all spread everywhere. It was impossible to give out food, clothes. Actually, it was a really big disaster."[51]

National authorities are not particularly enthusiastic when care materializes as a result of local initiatives. A consultant for the Ministry of Migration whom we interviewed in Athens stated that "our planning, the central planning, is to have dispersion and not concentration in one area." He continued, "We try to use provisional structures,"[52] suggesting a preference for instrumentalizing a reified infrastructure over relying on an unstable and fluctuating configuration of actors, things, and ideas. In any case, the specificity and locality of grassroots initiatives sits uneasily with the Greek authorities. The mayor of Chios explicitly told us that he did not coordinate his efforts with grassroots initiatives.[53] In fact, both grassroots organizations and NGOs were forced to register and get accredited, which then limits the associations they can form with other networks. Our interviewee from Starfish reflected: "Lately, it feels a little bit like the state tries to exclude us."[54]

In addition to differences in the mobility of networks, the circulation of things illustrates some of the characteristics of the humanitarian border. Things are not just objects; often they are inscribed with messages. In other cases, they exemplify the compromises that a humanitarian border consists of on a very material level. This focus on the circulation of things, technologies, and materialities has increasingly gained scholarly attention. The conversation among security studies, international relations, and science and technology studies approaches has, among many other things, resulted in a particular interest in the materiality of security and humanitarianism, as well as in the various ways that technologies aim to combine both care and control. This led to the identification of "humanitarian technologies" such as biometrics, genetically modified food aid and vaccination programs, and "non-human humanitarians" such as dogs, drones, and diagrams of particular forms of tents in refugee camps.[55]

In the case of this particular humanitarian border, specific things reveal different degrees of mobility. The commander of the Lesvian coast guard recounted how they "received vessels from other areas, as well as from Frontex."[56] Ships appear to circulate easily, as do personnel. He recalled how coast guard personnel staffed vessels on loan from NGOs, and personnel of the Hellenic Coast Guard staffed vessels borrowed from Frontex. But although vessels and personnel may durably circulate in various networks, Solidarity Kitchen, a grassroots organization that provides meals in the central park of Chios, depends on food and ingredients provided by residents and local suppliers.[57] The grassroots thus have a different relation to the objects that they move than the NGOs. The former relates to these objects in terms of reciprocity in communities, and the latter in terms of distribution. Objects thus also function differently as mediators in interactions with refugees. For the grassroots organizations, objects translate their aims of promoting mutual respect and equality, while for NGOs, objects are not attributed such symbolic functions. For their part, the Greek national authorities view grassroots initiatives as dysfunctional, as they divert control over provisions from the authorities themselves.

Simultaneously studying "the forms of circulation and the circulation of forms" also keeps us from too easily attributing deficits in policymaking and humanitarian support to chaos and disorganization.[58] Instead, studying the forms of circulation and the circulation of forms is likely to create awareness of the patchwork nature of border infrastructures and the central role of provisions and improvisations within them. Emphasizing this patchwork of interactions and transactions relativizes the importance of stabilized infrastructures. "We have the infrastructure, the materials, and the medications," stated the interviewee from Médicins du Monde.[59] But these infrastructures work only when coordination and cooperation keep people and things circulating on the inside. The mayor of Chios stated that "the main problem encountered by the municipality was the lack of a plan which could foresee what we should do and the lack of infrastructures,"[60] while the deputy head of the regional authority on Chios admitted: "I believe that we were surprised and didn't have the time to realize what happened . . . We didn't know who was responsible for doing what. We improvised at that period."[61] The regional authorities on Lesbos also emphasized the need to improvise, not due to lack of strategy but to lack of people: "We need personnel . . . we don't have personnel to staff the different committees."[62]

On Lesbos, a deputy head in the regional authority said that "there are too many NGOs on the island." "They help, but the situation is out of control. I have asked directions from the Ministry [of Migration] in order to have control over NGOs."[63] But for the general director of the police on Chios, it was not the NGOs that are hard to manage; it was "the independent solidarity people and volunteers" who are "the uncontrollable ones."[64] In counterpoint, a member of the grassroots organization Lathra pointed her finger at the NGOs: "Greece provided NGOs a huge area to play, not just to exist, but to play a central role, to create incidences, for good or for bad. In Lesbos the NGOs created a mess."[65] How different networks can more generally work at cross-purposes was reiterated by our interviewee at the Norwegian Refugee Council: "On Chios, you have oranges which are distributed to refugees for free and in parallel, you have people who don't want potable water to be delivered in Vial."[66]

If there is any truth to the saying that the border goes where migrants travel, infrastructures in the context of international mobility are shaped as much by the people on the move, the vehicles they use, the routes they take, and the people and organizations (from smugglers to banks) that support them as by agents of the state. Selection mechanisms do not only function at the border. The mediation of the borders continues. This can be illustrated by an account of researchers following the consequences of the legal and technological networks of the European Asylum Dactyloscopy Database (Eurodac) to Igoumenitsa, the last Greek port town before Albania. There, they visited an informal settlement of almost all male migrants, who have since been evicted by the police. They observed that "because many of these migrants carried the border even on their bodies (many of them already had registered fingerprints), they weren't able to completely cross the border that was literally embodied in the shape of their own fingers . . . they carried the border further themselves and, at the same time, transgressed it."[67]

Death, too, follows the border. As researches have noted, a tragic aspect of the material culture of border crossings concerns the crypts within trucks used to hide migrants. As noted by Galis, Tzokas, and Tympas (2016), "Compared with the high-tech devices used to detect them (. . . [such as] electronic thermal cameras that function as scanners of crypts), crypts represent a low technology." The study of migration and mobile borders is usually restricted to technologies used by governments to block access to Europe. To change this, materials need to be studied "not only in connection with the rhetoric of those who introduce it but also in light of how it is materialized in human bodies through clandestine border-crossing practices and material configurations (artifacts)."[68]

While border infrastructures often travel with migrants, not everything travels in the same direction or reaches the planned destination. This applies to humans as well as to things. Some things are left behind or lacking. An employee of Médecins du Monde on Chios told us that provisions were stolen from their shed almost every night. To protect these items, they hired a private security company that now protected not only the shed, but also those living in the camp.[69] The arrival of objects on the islands thus created new and expanding networks, some of which also were irritating. The former deputy head of the coast guard fumed about the garbage left on the island: "There were complaints by the residents and professionals in the port of Chios, mainly due to the issue of cleaning public space . . . They come here, we help them, we provide them with clothes and food, but they don't respect the place where they came."[70]

On Lesbos, garbage similarly became an issue, as abandoned life jackets were "painting red the shores of the 'red island.'"[71] Whereas for the coast guard, the abandoned objects were "garbage" and expressed disrespect, the meanings of such objects differed in the eyes of different publics. The coast guard's annoyance with the garbage left behind reveals yet another dimension of the relation between people and objects. Control over space is often central within disputes. On Chios, the activity of making meals for migrants in the central park has transformed it into a provisional camp, provoking reactions from villagers. A member of the Starfish Foundation on Chios told us that "during the summer, when more and more volunteers were coming, we start getting more and more criticism from the village."[72] A journalist on Chios pointed to the underlying dynamic: the more visible the "international presence" on the island, the more political the island becomes.[73]

Traveling Infrastructures

Chios and Lesbos are among the Aegean islands that received large numbers of migrants in the period of 2014–2016. The traveling of migrants and borders was explored in this chapter by describing the interaction and coordination—or lack of them—between Greek central and local authorities, international NGOs, the European Union, volunteers, local villagers, grassroots groups, and migrants. More specifically, this discussion described how border infrastructures—as a combination of humanitarian and security initiatives—can be said to travel. Rather than being the result of clear design, borders pop up at locations, being the result of the actions of various groups, organizations, and institutions, often with different, if not opposed, agendas.[74]

In these networks of social, political, material, and humanitarian tensions, borders may be said to travel with migrants. But although humanitarian border infrastructures involve all kinds of actors, there was no general planning or central overview. What our interviewees on Chios and Lesbos agreed on was the lack of any national or European government presence. According to our interviewee from Lathra, "the government left the NGOs to play their game."[75] This was confirmed by the mayor of Chios: "to a great extent, self-organization has covered the lack of state intervention and coordination."[76] A journalist working on Chios explained that "it was very convenient for them [the national government] that the local societies were self-organized at the beginning. It was our fault [referring to the local people on Chios] that we showed such willingness to deal with the problem. We sent a wrong message to the government; we showed that the issue is addressed on the islands. So, the government withdrew."[77] Another journalist working on Chios pointed to the gulf between the realities on the island and the place where decisions are taken. About the local authorities on Chios, he said:

> They are the authorities that everybody turns to when something is happening here, but they are also the authorities who are the least involved in creating all of this . . . Most of the decisions are taken elsewhere, and most of the information is also kept elsewhere. They are as much in the dark of the problem here, even more in the dark than the local volunteers. They seem to be just completely out of the loop.[78]

Although many grassroots volunteers were not shy about their ideological sympathies, both grassroots initiatives and NGOs tend not to see themselves as political actors. An employee of Médecins du Monde on Chios commented on her organization's stance toward the authorities: "I'm not sure if I can comment on this. I cannot say what the right thing to do is, and Médecins du Monde does not concern itself with policy issues."[79] A member of the Starfish Foundation on Lesbos told us that "we are born out of wanting to help here . . . basically, we've always seen ourselves as just helping and staying out of the political side of it."[80] This was also the understanding of a journalist working on Chios: "It is all very nonpolitical, and some people are really proud that they are not political. But refugees have increasingly asked us to focus on the political as the political situation deteriorates."[81]

The hotspot approach at Lesbos does not solely concern the blocking and obstructing of migrants by fencing them. Instead, a form of "containment through mobility" is at work, which aims to govern migration movements via channels and infrastructures.[82] The humanitarian border is not a case of spontaneous generation. The provision of care and control on the Aegean islands of Lesbos and Chios and the infrastructures that were required in 2014–2016 arose out of overlapping networks. Acts of humanitarianism and securitization necessitated moving various materialities and creating novel spaces for diffuse agentic capacities to emerge. To a certain extent, the arising networks can be seen as a border infrastructure that travels *with* the actors. But following only one category of actor does not generate a comprehensive picture of these configurations. The imperative to "take objects of security as the starting point rather than the end result of an act of securitization" proves useful, so long as objects and subjects are seen as parts of peramorphic mediations.[83] Following migrants, migrant organizations, volunteers, international NGOs, and a diversity of state officials simultaneously helps to prevent a state-centric perspective on borders, while considering humanitarian and security approaches in tandem gives migrants, movements, and materialities their due in terms of the composition of border infrastructures and the compromises that underpin them.

The humanitarian border can be considered a compromise among security, migration, and humanitarian approaches. Technopolitical compromises are proposed neither at the level of the exchange of political ideas nor at the level of the construction of objects and things. Technopolitical compromises occur in between. The previous analysis shows that compromises tend to be contagious. Once a connection is established among the different programs of care and control, this alliance is likely to be peramorphically reproduced in all kinds of practices and techniques. Meanwhile, compromises are vulnerable. Compromises with regard to the technopolitics of borders easily become compromised themselves, as the fate of the detention centers at Lesbos and Chios has shown. A long-lasting lack of proper housing, food, clothing, and medical care has turned the camps into daunting examples of vulnerability and lack of humanitarian aid. The intended mechanism of the EU-Turkey Statement, which would allow refugees with a valid asylum request to enter and also return migrants to the other side of the Aegean Sea, has come to a halt. Infrastructural compromises are similar to regular social and political compromises that they are easily contaminated and run the risk of becoming compromised. To detect the compromises that constitute border infrastructures, the morphological shape of the technopolitics of borders has to be unpacked so as to distinguish the relationships between the composing parts.

The analysis clarified that humanitarian reason is not translated fluently into humanitarian technologies. In that sense, this chapter has confirmed the classic lesson from science and technology studies—namely, that technologies are not mere instruments that express the intensions or the will of their makers without getting "lost in translation," so to speak. Various authors have described biometrics, vaccination programs, and tents in refugee camps as humanitarian technologies or "non-human humanitarians."[84]

The analysis in this chapter showed that tensions of humanitarian reason resonate in these humanitarian technologies. Food, shelter, and medical and legal services were provided in an ambivalent atmosphere where care and control are closely intermingled, and where the blending of security and humanitarianism via all kinds of things and technologies create novel compromises and novel compromised associations.

The humanitarian border arose from the materialities that were mobilized by actors engaged with care and control, in a continuous redistribution of roles and responsibilities. From the perspective of mediation, a humanitarian border can be conceived of as an entity that constantly undergoes shifts—translations of its composing elements into novel infrastructures. The particular morphological notion of technopolitics that was developed in the discussion between Sloterdijk and Latour conceives it as a world-making endeavor, a bubble-blowing and atmosphere-creating machinery. The inner radar work of that machine consists of all kinds of specific material and movable infrastructures. From time to time, the translations among actors, institutions, and technologies of all sorts lead to particular solidification points—"collectives" of humans and humans. The vibrant relations and interactions that shape a reality result in a much less dichotomous situation than one that can be explained in terms of subjects and objects.

The humanitarian border is a hybrid construction consisting of materialities and movements of all sorts. However, the solidification of a humanitarian border always comes with a proviso: underlying tensions, ongoing conflicts, new contestations, or the creation of competing entities can herald another hybridization of the situation. The engagements of actors concerned with care and control result in shifting border infrastructures. At certain points, a border infrastructure arises in which a clear delegation of security and humanitarian tasks takes place. At other moments, an intermingling of roles and functions is at stake. The creation of an entity, even a precarious and temporary one, is also the moment at which this entity can be deployed. Not only does an entity give a certain materiality to a particular configuration by momentarily "black-boxing" the underlying tensions and contestations, as an entity, it can also be put into motion by other entities and become an instrument itself.

However, this chapter has already showed that the technopolitical movements that transform a humanitarian border from a policy concept into a thing and a network, and vice versa, leave many gaps. Compromises between care and control do not result in seamless infrastructures; rather, they create tensions and gaps. These gaps are not only material and spatial—they also contain an aesthetic dimension that make some tensions visible, whereas others remain invisible. In what way can this interplay be understood? How does the aesthetic dimension relate to the material and spatial aspects of border infrastructures, and what are the consequences for the actors, institutions, and technologies involved?

Chapter 7 redirects the gaze toward various infrastructural investigations, actions, and projects run by artists, activists, and academics that look at the politics of border infrastructures. In doing so, the actors conducting these infrastructural investigations not only contest border infrastructures or denunciate their consequences, they also turn the technopolitics of borders into a public issue. As such, infrastructural investigations add yet another dimension to the technopolitics of borders: the media and mediations that create borders.

Bulgaria surveillance, September 2015.
Source: Henk Wildschut.

7   Infrastructural Investigations

The List

On November 9, 2017, the German newspaper *Der Tagesspiegel* published a list of all the people known to have lost their lives in their attempts to reach Europe between 1993 and May 29, 2017. The list was composed by a civil society network supported by United, a nongovernmental organization (NGO). Its appearance in a German newspaper on this date was no coincidence: November 9, 1938, is known as *Kristallnacht*, the "Night of Broken Glass," when a pogrom against Jews throughout Nazi Germany was carried out by paramilitary forces and German civilians. These dramatic events are remembered each year on November 9, the International Day against Fascism and Antisemitism. The list published by *Der Tagesspiegel* appeared as a forty-eight-page supplement. Distributed within 100,000 newspapers, the list covered over 33,000 documented refugee deaths. In 2018, the British newspaper *The Guardian* conducted a similar action. To mark World Refugee Day, it published the most recent version of this list on June 20, 2018.[1] It mentions 34,361 migrants known to have died.

"The list," as it is often called, is one of many initiatives to identify, recognize, and remember the people who lost their lives and to draw attention to the humanitarian dimensions of border control. Attention for migrants and refugees, the cruelties of the sea crossings, and the prevailing conditions in the detention centers and refugee camps has been abundant in Europe, especially since 2014. The previous chapters of this book on the Aegean contained several examples. Artistic and activist initiatives have included creative ways to offer support, as well as diverse forms of protest. Controversial attempts to raise awareness include Chinese artist Ai Weiwei posing as the drowned three-year-old Syrian boy Alan Kurdi, lying dead on the beach. Investigative journalism, such as CNN's undercover operation that revealed migrant slave auctions in Libya, have reported on the obscure aspects of mobility partnerships and all kinds of injustices related to migration.[2] Human rights organizations have reported on the violation of fundamental rights, such as by the pushback operations of Greek coast guards.[3] Many European museums and art galleries have devoted exhibitions to the subject, while academic attention to borders, migration, and refugees has flourished.

Of all the initiatives that somehow question border politics, this chapter explores yet another form of peramorphic mediation, namely, the way that borders and border infrastructures are represented and how these representations interrelate with border infrastructures themselves. The focus is especially on academic, artistic, and activist interventions, as well as various forms of investigative media coverage that generate not only awareness, but reveal specific insights, information, and visual representations that mark something about the workings of border infrastructures and their technopolitics. Border infrastructures reveal a peramorphic politics, in which political thoughts and actions are intertwined with the technological means to control borders and manage and monitor human mobility. Border infrastructures create specific political relationships, not only with the people crossing borders who are subject to monitoring, but with those who monitor the border and its technopolitics.

While political relationships come in many forms, the most familiar is the one between citizens and the state. Elections, taxes, bureaucracies, provisions ranging from health care to education, and various forms of registration testify to the organization of this political relationship. In studying border infrastructures, it is not only the relationship between the state and its inhabitants that matters, but between states and people on the move. Modern states have the authority to determine who can circulate within and cross their borders. They control "the legitimate means of movement."[4] States have always been bothered by people on the move, and this shows how modernist planning is imbued with a strong aesthetic component that seeks to make the behavior of citizens visible and entire societies "legible."[5]

The so-called state that prevails in these accounts has seen important transformations. Although the border infrastructures developed by the European

Union (EU) and its member-states can still be seen as instruments to control mobility, they perform this task in ways that have changed the notion of state control—as have many other transnational border infrastructures elsewhere in the world. Borders go where the movement is, and states have followed. In doing so, borders have assumed new tasks and left others behind. Border infrastructures organize a specific political relationship between state configurations and people on the move, just as the way in which they monitor mobility suggests how these infrastructures themselves can be conceived. The technopolitics apparent in these border infrastructures is not just the result of an instrumental relationship or political design. Instead, there is an intimate relationship between humans, technologies, and specific ways of seeing, looking, observing, and visualizing mobility.

The technopolitics of border infrastructures addressed by various migrant groups, their legal representatives, volunteers, artists, activists, and national and international NGOs concerns a wide variety of issues, including the humanitarian consequences of migration; the status of refugees; rights, membership, and questions of belonging; the effects of border controls on migrants; the politics of the European Union and its member-states and the justification of border controls; and the rise of violence, xenophobia, racism, populism, and nativism. Rather than studying this media attention or the social movements and artistic engagements that have addressed borders and migration, the analysis starts from the thesis that there is an intimate relation between border infrastructures and how their attendant political issues are made visible or remain invisible in the public sphere. Stated differently, the technopolitics of border infrastructures is not restricted to state actions; it includes the contestation and opposition that confronts it.

As the chapters up to this point have shown, border infrastructures are not just large-scale technological projects. They include numerous networks and information systems, as well as local and regional initiatives connected by interoperability, which often result in a "bricolage," a movable patchwork of border control activities. The previous chapters also revealed that the humanitarian border consists of all kinds of initiatives related to care and control, which result in an assemblage in which the construction and contestation of border infrastructures are hard to separate. For this reason, the aforementioned characteristics of border infrastructures apply to the actions of many NGOs, artists, and activists, as well as to the technopolitics of state agents. This is not to suggest that NGOs, artists, and activists constitute new governmental organizations that are part of the state apparatus or embedded in the greediness of the border regime. The presentations and representations of border infrastructures are often hard to distinguish, as their composition is the work of many hands and many eyes.

To explore these compositions, this chapter engages with infrastructural investigations—the various ways in which border politics and its consequences are studied and represented to the general public by investigative journalists, NGOs, artists, activists, and academics. Infrastructural investigations are a form of peramorphic mediation. Critical representations of border issues do not solely address human tragedy, the shortcomings of international politics, and the effects of existing border infrastructures. To a certain extent, they are also part of these border infrastructures, as they multiply the visual presentations and representations of border issues and contribute to their extensiveness. Infrastructural investigations pursue the mediating movements that are characteristic of the emergence border infrastructures.

Reporting Border Infrastructures

The notion of "infrastructural investigations" can be clarified by returning to the list published in *Der Tagesspiegel*. If a play on words could somehow be appropriate here, the list is definitely moving in its presentation of the almost unimaginable size of the migration-related human tragedy that has engulfed Europe and its neighboring regions in the past decades. However, the actual movements that led to this list are less obvious. Neither the movements of the people represented in this list nor the investigative and representational movements of the people who made the list are immediately visible. Meanwhile, the list published by *Der Tagesspiegel*—like many other circulating lists, infographics, and visualizations—advances a factual and causal claim and a moral argument that opens a certain space to evaluate existing border politics.

The list provided by United does not stand alone; it is part of a network of lists. Lists can be considered a technique of governance, a technique that displays ways of knowing, ways of registering, and ways of regulating.[6] Lists also *do* things; they have a performative capacity.[7] Meanwhile, lists like the one published by *Der Tagesspiegel* display an expressiveness that is limited and exhaustive at the same time.[8] Lists make things visible, but they also make issues invisible.[9] One way to throw light on the making of lists is to examine the disputes underlying the subject. Like the study of controversies, the study of disputes can reveal how facts become stabilized or contested and what alternative ways have been considered to represent a fact as a fact.

In 2015, the Human Costs of Border Control project published the Deaths at the Borders Database, an open-source collection of individualized information about people who have died trying to enter the southern European Union between 1990 and 2013, sourced from the death management systems of Spain, Gibraltar, Italy, Malta, and Greece.[10] In contrast to the list by United and other lists compiled using similar methodologies, such as the Fortress Europe blog, which lists news reports of those who died on their journey to the European Union, and the Missing Migrants Project of the International Organization for Migration (IOM), the Deaths at the Borders Database does not rely on data sourced from news media.[11] News media are unreliable sources of data. For instance, the media do not cover every shipwreck, and media attention to an event may increase, but it can also disappear. In addition, not every news item is equally detailed. Moreover, different media may cover shipwrecks and the discovery of unidentified corpses, and that may lead to overcounting.[12] But the Deaths at the Borders Database has its limitations too. The database records only bodies that have been registered juridically, and the ways of recording vary a great deal.[13]

The aim here is not to dive into the technicalities of deciding which methodology is preferable to count so-called border deaths. Neither will the debatable definitions of borders and border deaths deployed in these lists and databases be discussed in detail. Rather, the analysis wishes to emphasize that none of these lists and databases offers "a view from nowhere."[14] They should thus be seen as movements that simultaneously create specific subjects and objects, facts and spectators. This point can be stressed by looking at other representations of border issues and the kinds of interaction that they encourage between those who see and what they are going to see.[15] Representations of border issues are not restricted to the making of lists and the counting of casualties. As the techniques to visualize events develop, international organizations, activists, and migrants have applications at their disposal to register incidents and to blame and shame states that have the eyes to see irregular situations but refrain from acting.

It is thus not only states that multiply and intensify their ways of seeing and monitoring. The UNHCR, for instance, provides an operational portal that visualizes refugee situations.[16] According to the IOM, "non-State actors involved in supporting, facilitating or reporting on migration and mobility have been profoundly affected in a variety of ways. . . . Non-State actors are increasingly operating transnationally and their businesses and activities are much less confined by geography than ever before. As geography becomes less of an issue, migration processes are inevitably affected. . . . Non-State actors are responding to migration in innovative ways through the use of technology."[17]

Examples of such innovative ways to respond to migration through the use of technology include migrants' use of their own personal devices to connect to online mapping platforms such as Watch the Med[18] and its alarm phone[19] to monitor migrants' deaths and violations of migrants' rights at the maritime borders of the European Union.[20] To illustrate the dramatic events and to gather evidence of the atrocities, George Clooney and John Prendergast organized the launch in 2010 of a satellite[21] (now no longer operating) to document violent attacks, human displacement, and mass graves in Sudan. These projects have been criticized for offering no more information than was already available from other sources, such as the testimonies of victims.[22] The IOM itself is setting up a mobile phone application called MigApp, which will offer migrants access to information on the migration process and available services in destination countries.[23] Other examples include philanthropists, NGOs, and individuals supporting migrants by employing new technologies, including drones. For instance, the Migrant Offshore Aid Station, based in Malta, launched a mission in 2016 that included two drones to patrol Mediterranean waters using day- and night-sensitive optics to send back high-resolution images.[24]

Technologies of visualization are intertwined with border infrastructures. State agents may use the border as a spectacle to demonstrate their willingness to act and to perform their sovereignty. However, it is not just states that create such spectacles. Intergovernmental organizations, companies, NGOs, activists, and migrant groups compete to create images to raise awareness. Europe's migration situation itself, particularly the dramatic events in and around the Mediterranean, can be grasped as a border spectacle.[25] In such a spectacle, events and the representation of these events are interrelated. Border spectacles visualize the agency of states, migrants, and all kinds of civil society organizations. The result is a visual cacophony, a cacorama. The cacorama is not just a visual or representational spectacle, but a material-technological infrastructure that encourages the movement of various objects, including pictures and images. Such so-called border theater can convince the public of the state's vigor in tackling migration issues. But spectacles also fuel controversies and increase the risk of fraud and corruption. As a result, mobility is likely to become a security issue, fueled by crises and emergencies.[26]

Visualizations of mobility can lead to scandals, such as when governments fail to follow the law or use violence. In the widely reported Farmakonisi pushback case of January 20, 2014, eleven refugees—eight of them children—lost their lives when their boat capsized as it was being towed through rough waters. When lawyers for the victims requested information about the coordinates of the patrol boat at different times during the incident, no such information was said to exist. Yet, it is clear that such information is routinely registered using coastal positioning systems and radar technology. All that was initially handed over by the coast guard were handwritten logs by the patrol crew, supposedly read from the on-board navigation system. The commanding officer ordered that video cameras provided by Frontex—used to aid ground personnel to monitor situations and to later provide evidence—had been switched off.

In response to such suppression of data, activists and legal aid services to migrants urge people at sea to use mobile phones and other devices to track their whereabouts during crossings as proof of where they were, and at which times.[27] The event inspired theater director Anestis Azas to write and direct a play, *Case Farmakonisi or the Justice of the Water*, which was presented in Athens and the Epidaurus Festival in 2015. According to writer and theater director Zafiris Nikitas, "the performance creates a performative and investigative arc that starts with the incident, but then follows the legal and personal ramifications of the drowning. Who was to blame for the incident? How did the members of the coast guard involved react to what happened? What are the legal parameters that come into play in such occurrences?"[28]

Multiple university departments are involved in the construction of monitoring technologies and Big Data analysis that can be used to support ethical and legal claims. With its Science for Human Rights project, Amnesty International is stimulating the development of new tools such as satellite imagery for the purpose of monitoring human rights.[29] The Push Back Map documents and denounces pushback operations at the borders of the European Union.[30] The Eyewitness to Atrocities project allows people to document human rights violations by taking photos and videos with their smartphones.[31] Such countersurveillance technologies can be used by migrants to post information on databases about violations in order to hold states accountable and to stimulate public awareness. Examples include the Border Crossing Observatory in Australia,32 the Arizona OpenGIS Initiative for Deceased Migrants,[33] and The Migrants' Files[34] in Europe, a consortium of journalists from more than fifteen European countries coordinated by J++,[35] an international team of data journalism specialists.

Monitoring can also take the form of self-monitoring, such as by the "appification" of migration. Mobile phone technology connects migrants to networks of family, friends, humanitarian organizations, and smugglers, but on the other hand, it also connects smugglers to agents, officials, and their networks. Digital connectivity supports movements, as information, advice, and money can be shared. Moreover, the IOM reports that "real-time coverage of movements and operations enable migrants to access useful information on where, when and how to travel."[36] As a result, migrants create "digital passages" [37] with their smartphones, varying from "mobile homes"[38] to "digital diasporas."[39]

Monitoring does more than register reality and create connectivity. It results in a contest of competing images to prove the legality or illegality of actions. While states tend to deny the existence of pushback operations that seek to prevent migrants from entering territorial waters by towing them away, NGOs such as Amnesty International have many well-documented cases to cite. Like the detectives in the *CSI* TV series, researchers reconstruct the story of an event by applying all kinds of technologies in their quest for evidence. For example, the Forensic Architecture project reconstructed the journey of a boat that left Tripoli on the morning of March 27, 2011.[40] It ran out of fuel and was left to drift for fourteen days until it landed back on the Libyan coast with only nine of the seventy-two passengers surviving.[41] While the question of whether the states involved can be held legally accountable is still up in the air, the reconstruction shows that current mobility management resembles a situation of "organized irresponsibility."[42] The lack of cooperation between countries and the unwillingness to provide humanitarian aid result in situations where it is hard to blame a single actor.

For this reason, the kind of violence conducted under such circumstances can be typified as infrastructural violence.[43] The notion of "infrastructural violence" sharpens the relationship between border infrastructures, violence, and the agentic capacity of humans and nonhuman entities to create and execute such violence.[44] Notions such as "organized irresponsibility" and "infrastructural violence" reveal something about the technopolitics of border infrastructures. More than simply the material infrastructures that define the boundaries of state territories, border infrastructures are better understood as dispersed configurations that seek to organize the circulation of people. Of interest here is their ability to move and how they function as vehicles for political thought and action. The study of controversies often discloses how facts and knowledge are composed as well as contested. Likewise, the study of border controversies can help us to understand how border infrastructures are composed and contested and how they give birth to a specific form of technopolitics.

The Rise of the Observer

Border infrastructures can bring about infrastructural violence. But their other aspects deserve attention as well, such as their ability to encapsulate compromises, as outlined in chapters 4, 5, and 6. The initiatives described here have showcased how infrastructural investigations can be pursued. They shed light on the technopolitics of nonstate actors, especially their capacity to create counterinformation, counternarratives, and countervisualizations by mobilizing various kinds of information, research findings, mixed methodologies, visual techniques, websites, and technical formats. But just like state efforts to monitor migration, they do not achieve the "god trick of seeing everything from nowhere." Instead, they emerge from specific "situated knowledges."[45] These situated knowledges do not necessarily break with the idea of objectivity. Rather, their very situatedness underlines that the representation of objects and facts always involves the drawing of boundaries, practices of demarcation required to mark a specific perspective—even (or perhaps most evidently) when it aims at objectivity.[46] The representation of objects via visualization must be understood symmetrically, as subjects (the public of spectators able and entitled to read objects) are created together with the objects displayed in lists and graphs.[47] The representation of border issues is thus not only about collecting facts, but also about the creation of collective sight. In other words, objects and subjects are constituted simultaneously; the representation of border issues is accompanied by the construction of a particular way of seeing and a particular category of spectatorship. The following discussion continues the analysis of the movability of borders and border infrastructures by paying particular attention to the movements involved in both political and visual representations.

The question that will be explored next is how infrastructures and their representations are interrelated. This will happen by analyzing how representations reflect a certain object, but also constitute a subject, the spectator or observer who sees the representation. As argued at the outset of this chapter, border infrastructures create specific political relationships, not only with the people being monitored, but also with those who critically monitor the workings of the border. Border politics is a morphological endeavor. Borders have a certain size and shape and material extensiveness, but they are visual entities as well. Whereas the previous chapters engaged with border infrastructures surveilling migrants, I now turn to the political relationship between border infrastructures and the people who, in the words of the Israeli architect Eyal Weizman, initiator of the Forensic Architecture project, seek to reverse the forensic gaze to investigate state agencies.[48]

Theories of aesthetics also apply to ways of seeing in security policies, as well as to specific forms of protest in the context of border politics and detention centers.[49] The term "aesthetics" here refers not only to a theory of art or notions of beauty and the sublime, but also to a theory of observation and visual experience and the various visual ways in which technopolitical representations take place. Of particular importance for the purposes of this book is the notion of "lines of sight."[50] Applied to security and surveillance, lines of sight are "lines that segregate and divide, 'dividing practices' that render ways of life economic, make them amenable to management, trading, or exchange." They visualize and open up not only existing situations, but unknown futures, such as through algorithmic calculations. As Amoore says, algorithms "function as a means of directing and disciplining attention, focusing on specific points and cancelling out all other data, appearing to make it possible to translate probable associations between people or objects into actionable security decisions."[51]

Some theories of aesthetics that work well in the analysis of twenty-first-century border infrastructures originate from studying political relationships among people, technologies, and states in the nineteenth century. We thus need to rethink the state, and the aesthetic political relationships among border infrastructures, observers, and those being observed as states today are very different configurations from their nineteenth-century predecessors. But before emphasizing the differences, let me turn to the history of perception and the shifts that occurred in the late eighteenth and nineteenth centuries.

The transformation that took shape in the formation of states and citizens reveals something about how the public at large becomes involved in border infrastructures. Witness the changing interpretation of the concept *theatrum mundi* (the world as a stage). The view of the world as a stage comes from Greek and Roman times. The metaphor of the theater has been used extensively since the eighteenth century to describe democratic political institutions in bourgeois liberal societies. The metaphor has three elements: the stage on which the theater takes place, the actors who play on it, and the spectators who watch it.[52] While all three constitute the metaphor, the role of the spectators has been emphasized only since the eighteenth century. Although the spectators are offered a full view of the action, they typically remain invisible; at the same time, they are touched by what is made visible on the stage. They are involved but generally do not participate. This distinction between contemplation and action is a defining characteristic of the public sphere: citizens are free to move, to see, and to gain insight, but they are kept at a certain distance—a distance that enables them to create a view from *somewhere*. The subjective perspective of the viewer arose in the nineteenth century from developments not only in the arts, but also in society, politics, and technology. The type of viewer that emerged was the observer. The observer contrasts with the spectator: the observer is not a distant viewer, but rather a participant, conforming to implicit or explicit rules and conventions.

The emergence of this specific way of looking developed hand in hand with new technologies. Whereas the paradigmatic visual technology of the seventeenth and eighteenth centuries was the camera obscura, in the nineteenth century it was, among other things, the stereoscope. But it would be too simple and one-sided to tie the history of the viewer's role to the invention of new optical instruments, as these in turn are embedded in broader social and economic developments. The important development is not so much in the concrete mechanisms of looking and seeing, but in the changing abstractions of what looking and seeing *mean* in the reconsideration of what reality and realism encompass.

The observer in the nineteenth century emerges within a constellation of new events, forces, and institutions, loosely and perhaps tautologically definable as "modernity." "Modernization" here refers to a process that largely coincides with the spread of industrial capitalism and the mass movement and circulation of goods and services that were previously exchanged individually. The observer does not stand outside of this process as a spectator but is immanent in it as a human subject.[53]

But how did perception itself develop?[54] This is best exemplified in the development of photography, which has led, like no other technology, to an industrialization and commercialization of the production and distribution of images. Through photography, seeing acquires a more autonomous character and is empirically isolated from other sensory experiences. The visual emerged as the domain in which political relations take shape. The gaze that both states and citizens deploy in the nineteenth century takes a specific form. This is not the place for an exhaustive historical analysis of citizen as viewer. Nonetheless, an aesthetic political theory must investigate the relationship between state and nonstate actors in a dynamic way. Not only will this do justice to the role of citizens, journalists, NGOs, activists, and artists; it will show that the visual only emerges in the movement of looking and this movement is not one-way traffic. This analysis of seeing back and forth through citizens and states offers an additional argument to view configurations shaped by border infrastructures as political relationships.

The Detecting Eye

The watchful eye has transformed the spectator into an observer. As a result, the eye that watches the state and studies its continuously changing configurations has become a detecting eye. As recounted in chapter 2, controlling the common external borders of the European Union involves cooperation between EU member-states; investigating EU border infrastructures thus implies investigating transnational state infrastructures as well. Whereas numerous historical studies have tied the building of infrastructures to processes of state formation, many of today's border infrastructures are part of transnational constellations that include both states and a variety of nonstate actors ranging from NGOs to private security companies. Concepts such as "infrastructural Europe" underline that international political relations, cooperation, and diplomacy cannot be fully understood without considering their technological dimensions. Throughout this book, it has been emphasized that technologies can function as vehicles for international politics itself. How should the political relations constituted via technologies be conceived? In many cases, it is the ontological instead of the merely anthropological or instrumental relation that prevails in technopolitics. But what kind of ontological politics defines the political relationships among citizens, migrants, and border infrastructures? And which kinds of state configurations appear via these relationships?

The composition of border infrastructures can be examined from the perspective of a detective.[55] How does such an infrastructural investigation take place? An example is the investigation of an innovative transport system in Paris, dubbed Aramis. In this transport system, underground carriages can be coupled and uncoupled so that there is no longer a need to change trains. Aramis was ultimately not built, but why not? Bruno Latour's *Aramis or the Love of Technology* tries to unravel the plot.[56] Written in the form of a whodunnit, the book adopts the classic form of the duo, a wise and experienced teacher/detective (modeled after the author, Latour, as a kind of Sherlock Holmes) and a young student, eager to learn but somewhat naive, who together search for Aramis's murderer. The book can be read as empirical social science research into why a transport system failed to get off the ground, as well as a study of the interaction between concepts and reality. The philosophical detective investigates the transport system Aramis that never came into being. To investigate the matter, the teacher/detective sends the student off to open the black box. Black boxes, just like the information storage devices on aircraft, contain processes and practices solidified into facts. Anyone who opens a black box will see the social, cultural, and economic dynamics that had been locked away to give that fact its apparent solidity.

For Gilles Deleuze, a book of philosophy should be a specific type of detective novel, in which concepts intervene in reality and change in line with the problems. Underlying this is his idea about the nature of the reality to which philosophy must relate. According to Deleuze, the tradition of Western thinking is arborescent, much like a tree firmly rooted in the soil, branches sprouting from its trunk. In contrast to this image of solid ground under our feet, Deleuze proposes the rhizome, the rootstock that branches off in different directions, forming a network that spreads underground. Border infrastructures to a certain extent resemble this rhizomatic structure. This offers an opening to conceptualize the various investigations described earlier in this chapter as detecting investigations. This thesis can be supported by analyzing the development of the detective in crime fiction.

The development of the figure of the detective in literature not only reveals the genealogy of a specific genre, but also points to the relationship between state authority and the watchful eye. Analyzing the role of the detective is thus a way to shed light on the role of the state in ordering society. While historians of crime fiction disagree on when the detective novel originated and how far the genre extends, many books on the subject argue that it makes sense to talk of detective novels only since professionally organized police forces were set up in the nineteenth century.[57,58] The origin of the detective (and spy) novel is linked to the emergence of a specific form of suspicion related to state building at the time. The relationship between the state and the detective is crucial, with the detective novel emerging alongside psychiatry, political science, and sociology.[59]

The detective novel developed against the background of the shift from sovereign to disciplinary power in the late eighteenth and early nineteenth centuries.[60] The shift ushered in, among other things, a new attitude toward criminals: it was no longer sufficient to punish wrongdoers; they also had to be disciplined to improve. The shift to disciplinary power introduced new institutions such as prisons focused on rehabilitation, as well as new methods of administration and research, new ways of looking and organizing. As the eye of the state metamorphosed, so did the eye of the detective.

This transformation continued into the twentieth century. The transition of the traditional Victorian detective into his hardboiled successor is described by Deleuze, himself a fan of the genre, as a twofold change in style.[61] In a short piece to mark the one-thousandth issue of *La Série Noire*, Deleuze argued that the Victorian detective, of which Sherlock Holmes is the symbol, employed two forms of investigation: the French deductive method and the English inductive method. The hardboiled detective put an end to both of these traditions; stories in this vein make it clear that detective work has nothing to do with the scientific search for truth—it has everything to do with deception. Against the background of an untrustworthy and corrupt police force, what it means to solve a case also changes. It no longer amounts to solving a puzzle; increasingly, it is a question of tying up loose ends. The detective's involvement itself also changes as the case develops. In other words, the sleuth's interpretative and moral flexibility—as well as his affectivity (but not necessarily commitment)—grow.

The transformation of the relationship between the detective and the state leads us closer to an understanding of the nature of the political relationships constituted by border infrastructures. The classic detective in the work of Sir Arthur Conan Doyle solves a mystery. On the other hand, the hardboiled detective does more than solve a puzzle. The nature of the investigation shifts and adopts the aim of unmasking. The detective, however much he might be an outsider, is ultimately a defender of the existing social order as he sweeps chaos and uncertainty away from the representation of society.[62]

This is fundamentally different from the kind of detective work conducted by initiatives such as the Forensic Architecture project. It gradually becomes less clear to the detective how transparent evidence can be, whether situations can be correctly read and interpreted, and whether the trails followed will point to a culprit, motive, or cause. The Greek term *prosopopoeia* can refer to ways of writing and speaking in which objects can testify to stories and events.[63] The Victorian and the hardboiled detectives take a series of clues and reduce them to a clear explanation. They put the reader in a position to comprehend a meaning and a cause and to discover a motive. This changed radically in what is called postwar, postmodern detective fiction—a change that typifies the current political relationship between observers and technopolitical configurations.

The example par excellence of the contemporary detective can be found in *City of Glass* by Paul Auster, the first of three novels that later became known as *The New York Trilogy*. The detective in this story follows a man who walks along a particular route and, in doing so, sends a signal—but what does it say? Auster writes in *City of Glass*:

> The detective is the one who looks, who listens, who moves through this morass of objects and events in search of the thought, the idea that will pull all these things together and make sense of them. In effect, the writer and the detective are interchangeable. The reader sees the world through the detective's eye, experiencing the proliferation of its details as if for the first time. He has become awake to the things around him, as if they might speak to him, as if, because of the attentiveness he now brings to them, they might begin to carry a meaning other than the simple fact of their existence.[64]

The questions at stake here are very different from those of the classic detective novel or the hardboiled detective tale. What is the status of proof through the text? What does it mean? Is the proof direct, or is it mediated by something? Does it lead to anything? Does it form a clue? A clue to what? To a crime? To an event? To a particular pattern in reality? Or does the pattern point to something else? A relationship in which the detective himself is involved? And maybe, is it a red herring, something to throw us off the track?

Auster's *City of Glass* can be read epistemologically (as a reflection on what the term "method" or "investigation" means), semiotically or post-structurally (looking at how meaning is constructed and the role played by references). The text also lends itself to ontological analysis.[65] The route taken by the suspect Stillman creates, as it were, the perspective from which it must be viewed. The novel shows that orienting oneself within reality implies, to some extent, reorganizing it. It opens some cases and closes others, but either way, it makes the detective a part of it.

The Reconstructing Eye

The figure of the detective offers an opportunity to reflect on the relation between states, infrastructures, and events. Detection does not take place from a fixed point: it requires mobility and imagination. Two projects concerning border control in Europe illustrate this. The first is the Crossing the

Mediterranean by Boat project, which gathered the stories of migrants traveling to Athens, Berlin, Istanbul, and Rome.[66] An interactive story map allows users to read and to view the complex journeys of migrants. The second is the Migration Trail project, which displays fictional characters that set out their journeys from the shores of North Africa and Turkey.[67] Neither project simply tells a story of true or fictional characters; each also invites the observer to participate in the narrative by navigating through these stories and adding new ones.

The spectators act by observing, selecting, comparing, and interpreting; they relate what they see to all the other things that they have seen. Politics emerges when there is movement in what is perceptible, and when previously invisible elements of an established order become visible, thereby changing what is deemed commonplace. In this case, politics consists of creating new collective formulations by redefining what was previously taken for granted, by developing new ways of understanding the understandable, by giving rise to new configurations between the visible and the invisible, the perceptible and the imperceptible, and new distributions of space and time.[68] In these configurations, what becomes apparent is "the part that has no part."[69] Something that was kept outside the existing political order becomes explicitly manifest. The visual arises as the sum of power relations, social orders, maps, and images; it is not just an image that sprouts from a perfidious mind. The political order is constantly in a state of legitimizing itself as groups, people, and problems that are not part of the existing political order articulate, form themselves, serve themselves to be discounted, and cross the boundary between the invisible and the visible. The struggle for the political order and who or what is part of it is also an aesthetic struggle. It revolves around perception and visibility and is conducted through visual means.

The visual covers the entire domain of political visibility and invisibility, from slave plantations to the British Empire to the post–World War II military industrial complex. These are not just power systems with specific discourses. Instead, they can be considered as hegemonic visual systems that impose order. The visual is not only a form of representation; it intervenes in the making of orders and classifications that place groups and individuals within social and economic systems. Thus, it is a system of authority.[70] The emphasis on the visual should not be taken as a mere technical affair.

Visuality relates to questions concerning the legitimacy of state power. Legitimacy is expressed by visual means, as it requires techniques of representation to reveal the sources of its authority.

The detecting eye of civil society initiatives has largely been addressed by political theories that are less concerned with the sources of justification of state power than the possibilities of citizens to monitor and evaluate how it is exercised to make people, things, and events visible or invisible.[71] Contemporary citizens require a certain mode of detection to critically examine state policies. Political relations between the citizen and the state are only partially expressed in the representative organs of democracy; the eye is underestimated as a political organ, while a political theory that focuses on visual politics has much to teach us about the meaning of power. As Jeffrey Green says, "Popular empowerment under the ocular model does not involve the crystallization of the People's voice into an authoritative decision, but rather refers to the elevation of the People's spectatorship into the status of a gaze. It is the gaze—that hierarchical form of visualization that inspects, observes, and achieves surveillance—that functions as the chief organ of popular empowerment under the ocular model."[72] For the eyes of the people to see, eventfulness is crucial. Events create political moments and opportunities for people to hold governments accountable, as we saw in Forensic Architecture's investigation of the "left-to-die-boat." But eventfulness also has its drawbacks. An ocular perspective runs the risk of promoting a "politics of passivity."[73] When transparency becomes the most important quality of governance and management, the people are only left with the option to approve or reject decisions made elsewhere.[74] Watchful people are assigned a role in the forum, but institutions remain behind closed doors.

There is a particular relationship between the occurrence of border events and the kind of spectatorship that is—or is not—involved. Events can bind publics to political ideas and strategies, but they can also generate nonpublics. The contest between strategies and investigations to visualize events or to keep them disclosed creates the possibility of specific reconstructions of events. But how should such reconstructions be conceptualized? How do they relate to the particularities of border infrastructures, and what do they disclose about the involvement of state configurations and technopolitics?

The analysis given here suggests that the act of reconstruction can be compared to detective work.[75] The detective is an important metaphorical figure, but it has its drawbacks too. Walter Benjamin already gave a warning in 1938 with his claim that "in times of terror, when everyone is something of a conspirator, everybody will be in the position of having to play detective." Initiatives by citizens to map the humanitarian costs of border controls may turn them into detectives that also act as a kind of border guard. The rise of the detecting eye reflects the intermingling of the processes of securitization and responsibilization described in the previous chapter.[76]

Detectives begin their work only after a crime has been committed. The detective searches in an environment whose creation he has not experienced and, like Ahasuerus, wanders within structures that he has not designed. The detecting eye is a reconstructing eye. Reconstruction, fitting together the pieces of a puzzle to complete an image, is not just an evaluative activity; it always refers to a special event, a critical situation that requires attention. The notion of eventfulness gains currency once "turning the gaze against the state" opens the way to detecting gaps in state architectures and their unfunded responsibilities.[77] It is crucial to the act of reconstruction that the original circumstances are unknown, or at least inaccessible; situations must be reordered in order to make them rereadable. The event happened out of the public eye. Investigations are necessary to find the traces that will lead back to the event and to provide insights.

Reconstruction links control with imagination and vigilance with visuality. While reconstructing politics takes place after the event, this is not only due to the temporal chain of events or the lack of possibilities to influence them. The notion of "reconstruction" also suggests a conceptual discrepancy, in that politics—by using words and concepts performatively to encourage action and intervention—opens a gap between the fixed meaning of policies and how they subsequently work out in practice. While the political repertoires of border politics often turn to the existing transnational framework of states to arrive at authoritative interventions, border control primarily consists of the monitoring and management of mobility, and its political repertoires to catch up with events will likely evolve as well. Under these circumstances, a discrepancy emerges between concepts that demarcate boundaries, define territories, and set limits and notions that address the movement of people and the technologies, agencies, and institutions that must travel alongside them in their quest to control mobility.

Detecting the Infrastructural State

The borders of Europe have created a material, spatial, and conflictual landscape that fuses issues of mobility, security, and secrecy. The intermingling of a material-technological infrastructure with a visual repertoire by both state and nonstate actors has implications for the visibility or invisibility of borders. Border control and migration management enlist numerous technologies and, as we have seen, several European programs have sought to make their applications interoperable in order to improve the collecting and sharing of data, to include more institutions and organizations, and to increase situational awareness. But far from being a seamless web, the resulting networks can be seen as a chain of events—a series of political and technological openings that are somehow connected, and yet far removed from the stable and robust system of border surveillance that they often appear to be.

The analysis in this chapter has introduced another dimension of mediation. Mediation concerns not only the inner workings of border infrastructures, but also the observation, detection, and reconstruction of events related to these infrastructures. As such, mediation is also constitutive of publicization, the creation of publics and public eyes that perceive border politics and turn it into public events.

Borders express themselves at the intersections of territory, representation, and movement. They also intervene in this triangle by transforming its composing parts as border control technologies blur the distinctions between various forms of territory, affecting the very idea of territory by how they monitor, visualize, and represent movement. Technopolitics affects borders in many ways, including how borders relate to representation. Borders are rarely clear-cut representations of political will. Conversely, the visual technologies deployed to monitor and register the movements of people rarely result in uncomplicated representations of human mobility. Instead, the creation of images and visualizations requires infrastructural compositions and compromises of all sorts. The blending of border infrastructures with technologies of visualization results in the emergence of a cacorama.

The question if and how public engagement and media attention can catch up with the movability of borders can be answered only by considering the characteristics of today's border politics. The externalization of border control and the technological external dimension raises all kinds of questions, including the relationship between technopolitics and secrecy in which the key actors include national and European intelligence services, as well as illegitimate regimes, local clans, and leaders. A politics of events frustrates the way that border infrastructures can be evaluated politically. Border infrastructures not only consist of material and informational infrastructures that make up the border, but also relate to a public knowledge and media infrastructure that provides the public at large with information on the political and moral justifications of border policies, their intentions, effectiveness, and proportionality. Infrastructures are thus concerned with the knowledge needed to build and sustain themselves, the knowledge they generate about international mobility and critical situations, and the circulation of public knowledge that reflects on their own workings.

The deployment of information technologies, government cooperation with industry and professional bodies, and the externalization of border control affect the practice of mobility management in the European Union, the checks and balances within EU politics, and its evaluation of emergent technologies. Migrants, NGOs, artists, activists, and academics increasingly deploy all sorts of strategies to contest border control policies and the information given by states. In order to analyze the nature of these contestations, the ocular dimension requires attention once more (i.e., the way that issues are made visible to larger audiences to stimulate public debate). Examples include the visual representation of migration routes and refugee camps, the launch of interactive websites that invite the public to report casualties, online mapping platforms to monitor the deaths and violations of migrants' rights at the maritime borders of the European Union, and visual reconstructions of infrastructural violence, such as by the Forensic Architecture project.

While border infrastructures create events, they can also reveal events or turn apparent nonevents into public ones and nondisclosed nonpublics into eventual publics, such as a publicly visible group of people related to a critical border event. Infrastructural events that slip through the far-from-seamless webs underline the incompleteness of the efforts to represent and monitor human mobility. Events have the ability to shatter glass houses and to bring people, situations, and territories back into the picture. Infrastructural events reveal the composed nature of borders and allow the emergence of a particular kind of maneuvering space, the rise of an infrastructural state resulting from the myriad border infrastructures. On the other hand, a focus on events may lead to a politics of passivity. Eventfulness guides the public eye to events but leaves the institutions—and the technologies—behind closed doors.[78] A focus on infrastructural events should pay attention to agents, institutions, and technologies, and in particular to the way in which they collectively arise via the creation of border infrastructures. A border infrastructure is not a stabilized background or a solid network that organizes circulation. Instead, it is characterized by mediations and extensions; it consists of delegations, translations, transportations, and transformations of all sorts of knowledge, technics, decisions, and intentions. Infrastructural events provide an opportunity to detect the emergence of such configurations and the perapolitics involved.

Public campaigning, activist media attention, critical attention by NGOs and humanitarian organizations, and the interventions of artistic, academic, and activist initiatives have addressed all kinds of issues related to border politics, including the humanitarian drama on the Mediterranean, the condition of migrants in detention centers, the European Union's border policies, and the lack of cooperation among member-states. Various forms of border opposition have shown the ways in which issues are made visible or invisible, with the representation and visualization of border politics revealing the transportation and transformation of the border itself. The technopolitics in these cases consists of the interplay between the visible and the invisible in surveillance and countersurveillance, with the seeing involved always being from a certain point of view and embedded in specific material circumstances. The resulting sociotechnical networks that relate human and nonhuman interaction include chains of associations in which relations are constructed, information is circulated, and connections are made. Border surveillance consists of the combination of all kinds of local and regional networks by way of interoperability, in which information is circulated and carefully reconstructed into representations that create situational awareness and call or do not call for action. Mediation as mediatization and publicization concerns the perception not only of border infrastructures, but also of its inner constellation. Mediation as mediatization and publicization reveals something about the nature of border infrastructures. Countervisualizations likewise point to this dispersed and mediated structure of border infrastructures. The detecting eye may uncover numerous policy failures, humanitarian dramas, hidden agendas, scandals, neglected responsibilities, the abuse of power, and illegitimate or disproportional violence, but its most important epiphenomenological result, which the final chapter of this book will continue to examine, is the detection of the infrastructural configurations of which states and unions of states are a part.

Train station, Hungary, September 2015.
Source: Henk Wildschut.

8 Extreme Infrastructure

Marking Points

Curiously iconic in both their visibility and invisibility, Europe's borders continue to mark points of mobility and fixity and inclusion and exclusion. From the cliffs of Dover and the Mária Valéria Bridge that joins Hungary and Slovakia across the Danube River, to the almost unnoticeable change in road surface between Italy and France; and from the camps on the Greek Aegean Islands, the fortifications on the boundaries of Hungary and Austria, and the migrants stuck in Ventimiglia and Calais to the cooperation between the European Union (EU) and the Libyan coast guard and refugees drowned in the Mediterranean Sea: Europe's borders arise and move, surveil and intervene, perish and continue in other guises. Borders are not only avatars of politics or instruments that carry the burdens of history and the Westphalian past that can be used at will; they also translate and mediate politics by creating moments where the conditions of territory are reproduced. Tools, devices, and instruments introduced to address specific challenges become parts of networks that quickly morph into border infrastructures. Witness, for example, the fingerprinting machines in Aegean reception centers connected to the databases of the European Asylum Dactyloscopy Database (Eurodac) and asylum applications under the Dublin Regulation. The growth of borders as infrastructure out of networked actors, institutions, and technologies entails the emergence of a machinery of governing and decision-making—an archive of images and imaginaries of past, present, and future movements of people, as well as of possible interventions. While technopolitics works as a web spun over political actors and agencies, the unwieldy technologies of border control are hard to steer and difficult to govern, manage, and coordinate. They come to impose their forms on the political functions that they ought to execute.

How has the changing nature of borders and border control affected Europe's development as a political entity? What forms of technopolitics are implicated in recent developments? In the previous chapters of this book, the gulf between "ontic" and "ontological" conceptions of technology was filled with technologies combining materiality and movement, while concepts such as "kinopolitics" and "viapolitics" pointed to their dizzying variety. To unravel the specific technopolitical nature of these movements, I suggested an approach that can attend to changes in the form of both technology and politics. Borders can be objects or instruments of state power, networks to organize international human mobility or worldviews to order reality. A technopolitical account of borders addresses the transformations between these repertoires and follows how technologies travel from one form to the next.

This chapter offers a final reflection on the technopolitics of Europe's border infrastructures. I first revisit the idea of borders as infrastructure, using the technopolitical movements revealed in the investigation of airports, surveillance systems, hotspots, and humanitarian borders to interrogate the particular peramorphic mediations of which they consist. I then argue that these mediations all point in a specific direction—to what can be called "extreme infrastructure." The book ends with a coda on the significance of COVID-19 for Europe's borders-as-infrastructures.

Peramorphic Mediations

Borders illustrate once more that politics and technology are inseparable. States require techniques to govern societies, which manifest themselves through concrete measures to collect taxes, keep census records, map populations to implement health policies, and much else. Border control fits into this list; like other technopolitical systems, its technologies require combinations of administration, organization, and coordination. While borders can be seen as infrastructural systems, their ability to control requires other supporting infrastructures such as surveillance networks, monitoring mechanisms, databases, and the Internet. With these technologies, states create new spaces of governance and new ways to monitor target groups, as well as new contact zones with other countries that are rarely free of tension.

The previous chapters characterized the technopolitical expansion of Europe's border infrastructures as instances of peramorphic mediation— the expansion of border infrastructures through the creation of borders and border practices, intensified relations between these practices, and the transformation of existing practices into new border infrastructures—all giving rise to multiple political relationships. The extension and multiplication of borders pose particular risks in the current era of growing violence, xenophobia, racism, and nativist populism, as borders become laboratories of movement where experiments are conducted on humans, affecting their rights and their very lives. I pursued the peramorphic politics of Europe's border infrastructures through the notion of mediation and by following the emergence of infrastructures—not as a flat ontology in which human and nonhuman are equals, but as a ménage à trois among actors, institutions, and technologies. The notion of technology itself consists of all kinds of mediations, encompassing mutually interacting things, networks, and worldviews. Mediations were also found in the infrastructural competition over the design of the European Union and of the mechanisms to mediate between Europe's internal and external borders. Mediation also concerns the infrastructural imaginations that situate Europe as a security actor and the infrastructural compromises that are required to do so.

In chapter 2, I discussed four characteristics of borders that make them comparable to other kinds of infrastructure. Although the border as infrastructure is comparable to other infrastructures, what intrigues me are the specific technopolitical dimensions of the border. In what follows, I revisit the infrastructural characteristics of borders. But what I previously termed as "characteristics," I now frame as "movements"—the back-and-forth transformations of the border as tool and instrument, network and worldview.

The humanitarian border exemplifies such movement. Once established as a concept, it has become a portable instrument that can be set up where necessary, operating as a pneumatic infrastructure that can be inflated, set up, and taken to a tracking location.[1] But between the humanitarian border as a policy concept and a portable containment tool lies its conception as infrastructure—one inextricably linked to various state and nonstate organizations, information systems, and human bodies harboring distinct fingerprints. Humanitarian borders appear in many guises, changing from one form to the other and giving rise to a morphological kind of technopolitics.

This perspective allows us to conceive of border infrastructures as particular peramorphic mediations.

The first peramorphic mediation concerns the relation between materiality and spatial movement. The infrastructural bordering of Europe is not restricted to the boundaries of the European Union or its member-states; nor is it necessarily steered from the center. Governing, securing, aiding, monitoring, and registering are all closely connected, relational arrangements. While the reintroduction of border controls in the Schengen Area, the opening of hotspots and detention centers in Greece and Italy, and the rise of walls in Europe during the migrant crisis of 2014–2016 may all seem exceptional, they can also be seen as more mundane effects of coordination between Europe's internal and external borders. Although terrorist attacks in Europe that occurred following the 9/11 attacks in the United States and the war in Syria have led to the intermingling of border control, migration, and security policies, the externalization of border control and the intensification of security mechanisms were set in motion long before. In the meantime, Schengen has begun experimenting with programs, policies, and technologies that bring together the controlling of its internal and external borders. The mechanism is much more than an instrument in the hands of EU institutions, or a tool to be used by Frontex; instead, it creates a wheelwork through which Europe is built and rebuilt.

The power, control, surveillance, and intervention that accompany the proliferation and dispersion of borders affect the relations between centralized and decentralized authorities, internal and external borders, and isolation and circulation. The dichotomy between the European Union's internal and external borders, between Europe's inside and outside, is being infrastructurally replaced by more dynamic relations that result in internal isolation (detention centers in EU member-states) and external circulation (the EU-Turkey Statement), organized through numerous compromises.[2] Borders are not only facilities to organize movement or assemblages that connect technologies, but also are deeply implicated in circulation.[3] Border control is organized to surveil human mobility to facilitate intervention. While blocking movement or preventing people from entering expresses such control, borders also enable the following of people through monitoring and registration, in effect checking them before they reach critical crossing points. Borders act and select, not only at the boundaries of states and territories, but also at numerous places where state authority, technology, and movement meet. The border as infrastructure is rooted in the design of both large-scale policies (Schengen, Dublin) and technological systems [Eurodac, Visa Information System (VIS), Schengen Information System (SIS)], as well as in more mundane elements. A peramorphic perspective on the items used by migrants shows that shipwrecks, tents, garbage, and life jackets littering the shores of the Aegean Islands constitute novel material openings for following the emergence of infrastructures.

The second mediation concerns movements to select different sorts of migrants. Border infrastructures distinguish migrants from travelers in specific ways, organizing circulation to identify persons who are not allowed access under certain conditions. Classifying migrants through categories such as regular and irregular is but one example. Such classifications give rise to mobility regimes that distinguish between insiders who belong and variously othered outsiders who do not. By following the organization of passenger flows, the control of travelers and the management of migration, we saw how relations between design, detection, and detention at the airport are expressed through infrastructural improvisations, innovations, and compromises. Border regimes stem not only from restrictive state policies, but also from infrastructural compromises that connect and disconnect various forms of selection, isolation, and circulation. To understand the inner workings of a border as infrastructure, we need a different scheme than simple inclusion/exclusion. We can see how the sorting function of information technologies and how they create classifications of various kinds of migrants—students, workers, temporary migrants, refugees—is the outcome of a sociotechnical process.[4] Rather than being two extremes in a fixed scheme, inclusion and exclusion apply to specific moments and locations where the categorization of persons comes to be associated with more, less, or unclear and ambiguous rights. Databases and registration systems mediate between existing oppositions such as between citizens and migrants and generate a proliferation of different kinds of people with different statuses and opportunities. As a result, the insider/outsider distinction is being replaced by a much more heterogeneous handling of technologically constructed categories of people.

The third mediation concerns the aesthetic interplay of the visible and the invisible. The intermingling of migration and security policies, Europe's repositioning as a geopolitical actor, and the border policies of EU member-states and other European countries have focused our attention on borders since the beginning of the so-called migrant crisis. The hidden integration of Europe through its infrastructural projects is now fully exposed.[5] But these infrastructures were never completely invisible; they were out of sight only to scholars who studied European integration and cooperation through treaties and documents while ignoring the material constructions of Europe. The idea that infrastructures are part of a stable background that structures society—which become visible only when technologies stop doing what they are supposed to do—has made way to a more variegated view. Infrastructures are visible and invisible at the same time; some parts are on the surface, while others are underground. Visibility is not something that can just be switched on and off; much depends on the event at hand and the specific constellation of actors, institutions, and technologies. Borders can be visible or invisible while harboring a particular aesthetic relation between vision and action. The European Border Surveillance Program (EUROSUR) aims to create situational awareness by visualizing specific events in order to intervene, whereas the hotspot approach operating within this program processes visual biometric information to assess applications while keeping the bodies of migrants and their living conditions in the camps out of sight. Monitoring borders implies constant zooming in and out, arriving at positions that allow only partial oversight while preparing for interventions in real time. Conversely, journalists, academics, nongovernmental organizations (NGOs), migrant support groups, and activists aim to make things visible by shining light on precarious situations, critical events, and the networks within which power and responsibility do (or do not) circulate.

The fourth mediation concerns mobility itself. While borders are concerned with movement, border infrastructures themselves are movable entities. This peramorphic mobility concerns EU and member-state policies, as well as the actions of migrants. The European Border Surveillance Program EUROSUR depends on the mobility of patrol boats and surveillance instruments, as well as the flow of data and transfers of knowledge, regulations, and policies; the humanitarian border entails the traveling and intertwining of care and control; the investigation of border infrastructures requires the mobilization of representations by architectures of knowledge and visualizations that reconstruct infrastructural events. These movements are not just instrumental; the externalization of border control and the rise of so-called mobility partnerships are a technopolitical worldview as much as a sociotechnical project.

Seen peramorphically, infrastructural imaginations and infrastructural violence travel hand in hand. And as we saw in the previous chapters, borders travel with migrants. Corridors of migration, migrants gathering for collective travel, and the goods, devices, and technologies used for housing, food, and medical care, as well as for communicating and conducting financial transactions all give rise to portable infrastructures that combine care and control. This transformation of borders into movable border infrastructures leads to a multiplication of political relationships. Local volunteers aiding migrants, professionals from international NGOs, coast guards, Frontex officials, municipalities in border areas, local and regional police, and the border guards of various countries all find themselves negotiating the often-conflicting demands of humanitarian support, border control, state politics, and the governance of international mobility.

These four mediations intervene inside and outside of states and the European Union in multiple ways. As infrastructures, borders are the technologically designed, delegated, mediated, and morphological manifestations of the circulation and selection of people and the control, anticipation and evaluation of their movements. They function through the use of myriad instruments, from walls and barriers to registries and Big Data. Rather than resulting in an all-encompassing surveillance system, we see the becoming of a fragmented field of border practices in which configurations of actors, institutions, and technologies are involved in the bordering of Europe.

Traveling to the Limits

The mediations that typify Europe's borders are not just the characteristics of a particular kind of infrastructure; they also point in a specific direction—namely, at the way that Europe's borders develop *as* infrastructure. This direction is toward what I call "extreme infrastructure." The terms that I have used to typify Europe's border infrastructures—"compromise," "mosaic," "patchwork," and "bricolage," to name but a few—underline the interoperability that is expected to connect the various extant border control systems. EUROSUR's goal of achieving situational awareness is a means to bridge the gap between vision and action to enable critical border interventions in real time. The humanitarian border aims to both protect borders and human lives. The airport's design can be seen as a specific architecture to facilitate intervention, whether to manage the flow of passengers or to detect and detain unwanted persons. Conversely, the infrastructural investigations described in chapter 7 can be likened to countersurveillance, deploying the "detecting eye" to reveal the contours of the infrastructural state and cases of infrastructural violence in order to raise awareness and create possibilities for public intervention.

These infrastructural compromises are more than the dialectical outcome of the open-closed dichotomy; the interplay between opening and closing reveals a morphological logic, a tension between political thinking and acting in terms of ends and endings (an eschatological view) and a generative view that emphasizes the birth and subsequent transformation of events. In the latter, borders do not mark a clear end or ending, but rather a multiplication and hybridization of bordering practices that can affect relations among states, peoples, territories, and technologies. The multiplication of borders concerns both the expansion of borders and the people engaged with them: migrants as well as rescue workers and volunteers who are increasingly criminalized for offering food and shelter. As infrastructures, borders never end.

This reconceptualization of infrastructure points to how Europe, as an infrastructural imagination, has been constructed and deconstructed through its border politics. As a boundary project, Europe and its border politics emerge at the threshold of actions and interventions by a myriad of actors, institutions, and technologies.[6] Although the political and institutional order of the European Union may be unique, its border infrastructures share many features with border infrastructures elsewhere, with comparable constructions, conflicts, and compromises. Rather than suggesting European exceptionalism, the notion of a boundary project allows us to situate Europe's border politics in the continent's particular regional environments, histories, and political orders in such a way that the politics remains open to comparison with other border infrastructures. Characterizing Europe as a boundary project, furthermore, acknowledges how Europe has developed via its border politics. Like the etymological origin of the word "project," Europe's border politics are thrown forth (which is what the Latin *pro-iacere* means).

If Europe is indeed a boundary project, what does it bring forth? It is tempting to see the development of border infrastructures as being guided by an infrastructural kind of state governance.[7] Indeed, some aspects of the planning, funding, and regulation needed to install border controls are comparable to the nineteenth- and twentieth-century expressions of the infrastructural state. But the current organization, coordination, and integration of state policies through technology is much more than an instrumental operation, as it does not leave the actors and institutions involved untouched.

I therefore suggest that we need to revisit our ideas about the infrastructural state. If technologies are not only means to an end but mediators of techniques of governance, is it not plausible that state authority not only is transferred or delegated to technologies, but also emerges from these very techniques? Europe as an infrastructural state arises out of the intermingling of policies, agencies, and technologies; its border infrastructures do not arise solely out of human design or political will. While controlling its external borders was one of the European Union's tasks from the outset, the legal order that developed around its policies of border surveillance did not provide a blueprint for installing border infrastructures. The European Union and the states participating in the control of its external borders developed from the inside out, from the very construction of border infrastructures.

The question of what Europe as a boundary project brings forth can be raised even more profoundly. Considering the fire that destroyed the Moria camp on the island of Lesbos September 8, 2020, pushback operations at sea, the deaths in the Mediterranean, and the conditions under which migrants who are part of mixed movements travel across the continent, Europe's borders often lead to uncertainty, insecurity, vulnerability, and violence. In these cases, Europe's borders have become the machinery of extreme situations.[8]

By "extreme," I not only mean how Europe's borders create states of exception or extreme increases in migration, often referred to as peaks and climaxes, waves and tsunamis. Rather, what I refer to as "extreme" materializes in how selections are made. Border infrastructures push inclusion and exclusion to the extreme by pursuing selection under circumstances that inevitably introduce states of exception, in which people's rights are endangered and migrants are reduced to objects in the politics of circulation. By technologically naturalizing processes of bordering, borders as infrastructure fashion the exceptional and turn it into the new normal. Increasing migration is often said to exert extreme pressures on Europe's borders. But human migration is not an extraordinary event; the tensions that accompany border infrastructures are "intra-ordinary" events, generated and disseminated by the dispersion and diffusion of borders, their movability, and the interplay between visibility and invisibility.

My conception of borders as extreme infrastructures is also informed by the rise of so-called gray zones and black holes where the control of borders crosses the boundaries of existing policies, regulations, and fundamental rights or leads to situations of organized irresponsibility and infrastructural violence. The technopolitical border policy of surveillance in the Mediterranean, the hotspot policy in the Greek Islands, and the outsourcing of border controls to strongmen in Turkey and warlords in Libya (to cite but two examples) are a recipe for compromising compromises.

The point of the notion of extreme infrastructure is that it not only encounters critical situations, but also to some extent creates exceptions through the way that the infrastructure itself is organized and operates by its peramorphical multiplications. In this sense, the infrastructure generates extremes. The movements are not simply from normal to extreme to the becoming of a new normal—extreme situations are not only exceptions but intensifications of normal situations. The etymology of "extreme" is apposite here, as the word is the Latin superlative of *exter*, meaning "on the outside," "outer," or "external." The emergence of extreme situations is not a development in which something that is inherently political (borders) slowly or rapidly radicalizes into extreme infrastructures. Following the technopolitical repertoire, the emergence of extreme infrastructure takes place through all kinds of movements that transform it from instruments to networks and vice versa. These technopolitical travels do not necessarily create extraordinary situations, but account for intra-ordinary maneuvers that make up all kinds of compromised compromises, resulting in a diffusion of extremes. As a result, border policies that are operative at the outside of the European Union (e.g., at the external borders), such as the externalization of border controls, also affect the inside via an internalization of border policies and technopolitical imaginations of borders. The ongoing multiplication, transformation, and hybridization of borders turns Europe into an infrastructural state with movable borders that organize mobility. Europe's current relationship with borders renders borders—and Europe itself—as an extreme infrastructure obsessed with boundaries and limits.

Borders are archetypical political entities, among the oldest objects of state concern. Whether they are perimeters to mark territory, mechanisms to manage migration, signs or symbols of national identity, or security instruments to control the mobility of populations, or take the form of theater and spectacle, borders are political. But borders are not just objects of political thought and action; they also are the things, concepts, and issues through which politics and states develop. As migration policies, security policies, and foreign affairs overlap, borders increasingly become the realm of politics; the repertoire of databases, checkpoints, registers, and monitoring instruments, as well as the linkages among border infrastructures on land, at sea, and in the air, guide international political cooperation. Politics give and receive shape via borders.

The shaping and reshaping of borders bring back ghosts of the past, histories of discovery, trade, conquest, colonialism, and racism. But they also foreshadow a future in which the bordering and rebordering of Europe will only grow. Besides offering a way for states—and in the case of the European Union, a union of states—to reinvent themselves, borders create an atmosphere filled with tension. The technopolitical nature of these tensions becomes apparent only when borders are no longer solely seen as representations and manifestations of state power, but as infrastructures that create spaces for technopolitical maneuvering. Checkpoints, walls, cameras, biometric systems, cell phones, registers, and databanks not only organize the movements of people, adding an informational layer onto the sociotechnical organization of states; they also compose an architectural configuration that facilitates intervention. The entanglement of technologies with landscapes and seascapes, as well as the intertwining of programs to provide traveler services, care, and control, results in a peramorphic multiplication of political relationships.

The infrastructural lens this book applied was partly introduced as an alternative for the focus on institutions. In the case of borders, the infrastructural perspective shows borders are not just the material realization of institutional policies and politics, but have a particular material dynamic that fuels their expansion and transforms their shape. As borders change shape, they transform from objects into networks into world views and vice versa. These peramorphical mediations (i.e., the infrastructural expansion, continuation, multiplication, and transformation of borders) complicate their evaluation. When borders are stretched to the extreme and split apart into all capillaries of societies, questions regarding their infrastructural legitimacy and infrastructural illegitimacy become more and more pertinent. As politics and political compromises morph into material shapes, addressing these questions does not become any less complicated. Meanwhile, these complications offer a ticket to follow the emergence of borders as movable infrastructures. As such, investigating borders as infrastructure is a next step to detect the unfolding of technopolitics in policies concerning international mobility and security.

As vignettes of the efforts to control human mobility, borders express the tensions among state authority, technology, and movement. The infrastructural bordering of Europe is likely to mingle with geographies and ecologies that affect the circulation of human and nonhuman entities. As such, borders continue to be the entities par excellence through which state configurations present and reinvent themselves regarding such issues as the consequences of climate change or the COVID-19 pandemic that currently grips the world. Climate change will affect border infrastructures not only through the growing risk of disasters, displacements, and environmental migration, but through the intensified circulation of nonhuman species caused by changing climates and the traveling of pathogens such as viruses, bacteria, and parasites. Borders may be dispersed, transformed, and displaced, but as movable entities, they are likely to persist as mediators of technopolitical interventions—and to guide politics well into the future.

Coda

Bordering the Pandemic

If journalism is the first draft of history, where does this leave research into the infrastructural status of borders during the COVID-19 pandemic? In the first half of 2020 the world was ravaged by the novel coronavirus. Currently, during the conception of this chapter, countless countries, economies, and societies still remain in lockdown during the second wave of the virus despite so-called exit strategies gradually being introduced. The dangers of the pandemic are far from over—without a widely available vaccine, new outbreaks await us. The development and distribution of a vaccine will likely be affected by struggles over power, profits, and national prestige that are already evident in the fight against the virus.[1] The pandemic will likely lead to other crises, economic, social, political, and geopolitical, and to a protracted wave of public health problems. For the foreseeable future, nothing seems unaffected by COVID-19.

This postscript addresses the significance of COVID-19 for borders as infrastructure, but with strict caveats. Although reporting on the pandemic's significance for European borders remains a perilous endeavor, the virus cannot be ignored—not only due to the nature of the health crisis and its undeniable consequences for borders and border control, but because the pandemic is revealing how easily borders as infrastructure slip into a state of emergency preparedness. Following my notion of "extreme infrastructure," the encounter of borders with COVID-19 is not just an extraordinary event; it also intensifies the possibilities of extreme infrastructural policies made possible by existing technologies. The pandemic thus presents not only an extraordinary situation, but an intra-ordinary event for a mobile infrastructure well versed in the art of targeted intervention.

Renewed border checks were among the first measures taken in Europe to contain the virus. The Schengen Borders Code allows member-states to reintroduce temporary border controls when public policy or internal security is seriously threatened.[2] The closure of land, sea, and air borders has duly restricted free travel in the Schengen Area, turning it into "a tangle of unilateral border closures, bilateral tourism agreements, and free-movement bubbles."[3] And although the word "pandemic" etymologically refers to all (Gr., *pan*) people (Gr., *demos*), the consequences of the health crisis are unequally distributed. So, too, are the consequences of closing Europe's internal and external borders. The COVID-19 pandemic is intensifying the infrastructural isolation, vulnerability, and often-violent living conditions of people in refugee camps and detention centers, both inside and outside Europe. These refugees often lack medical care, including access to masks, and access to clean water and nutrition, and they find it impossible to practice social or physical distancing. Further, travel restrictions are undermining the right to asylum of people fleeing war and persecution who seek refuge in European countries.

While the pandemic is drawing our attention to the links between border security, health security, and migration, what do mobile borders have to do with COVID-19? Human beings are not only the carriers of SARS-CoV-2, the virus that causes the disease; in their movements, they carry the border with them. Social and physical distancing implicates the reproduction and multiplication of borders as they penetrate society: public spheres are geometrically redesigned into public health spheres, while alternative routes and maps are rearranging the logistics and geography of social life. Discriminatory regulations are no longer imaginary; age and vulnerability, caused by, for instance, diabetes and lung diseases, become potential signs of contamination. As borders penetrate public life, they impose on bodies as well; in the era of COVID-19, the body is not just a fragile immune system, but a source of information. Vaccination certificates and immunity passports may well become the new tickets to mobility. The idea of the border is again incorporated in an infrastructure of monitoring, administration, and registration, linking the possibility of national and international movement to the health of human bodies and the risk of spreading the virus.

Another comparison that arises in the context of the pandemic is the technopolitical relationship among border infrastructures, the monitoring of international mobility, and systems of surveillance. Lockdowns in various countries and strategies to evade them are informed by modeling of the virus's transmission, including patterns of human mobility. Before

COVID-19, health security systems already included surveillance networks featuring epidemic intelligence (EI)[4]—namely, detection and containment policies and a technological apparatus consisting of crisis rooms, monitoring systems, web scanning tools, and early warning and rapid response systems.[5] A special form of health surveillance consists of Internet-based algorithmic syndromic surveillance.[6] The information technology tools specific to the European Union (EU) were largely made possible by the establishment of the European Center for Disease Prevention and Control (ECDC) in 2004. By working closely with the World Health Organization, the ECDC would "soon become a regional hub" and "a mediator in expanding a closer regional health security assembly around EI priorities."[7] Its tools and monitoring activities include the EU Early Warning and Rapid Response System (EWRS), the Medical Intelligence System (MedISys), and the Healthcare Effectiveness Data and Information Set (HEDIS).[8] Health security and surveillance seem to be following trajectories familiar to the development of border infrastructures, with representing and intervening going hand in hand.

While current efforts to improve surveillance are informed by public health concerns, the desirability of establishing an international system seems debatable when the exchange of information, the timely alert of outbreaks, and the ability to learn from the successes and failures of other countries are undermined by mistrust, cultural bias, economic strategies, and geopolitical gamesmanship. Like border surveillance policies to monitor and prevent migration, information networks and technologies of visualization lack accurate knowledge of the movements of people, patterns of circulation, and forms of human interaction. COVID-19 also suggests that surveillance is not the sole prerogative of states, as other types of knowledge seem indispensable for combating the virus.

Contrary to what many suspect today, "surveillance" in its original sense did not mean a type of supervision or the disciplining of citizens as in Jeremy Bentham's panopticon or Michel Foucault's population management. Instead, the term referred to the countercontrol of politics by the public.[9] When states and international organizations fail to report in a timely manner, share data, organize knowledge transfers, and gain public trust, the translation of epidemic models and the monitoring of human mobility into policy measures are doomed to failure. When states ignore scientific knowledge or refuse to benefit from the data that the public can provide—through either tracking and tracing or widespread testing—they miss the opportunity to benefit from the wisdom of the citizens. Here, I do not consider the public as an existing and already available reservoir of knowledge but rather, as outlined in chapter 7, as an observing, acting, and detecting infrastructure that carries out counterinvestigations and produces counterrepresentations that visualize what is invisible or what has been made invisible—and that compensates for the intended or unintended blindness of states.[10]

The integration of border and health security infrastructures again reveals the technopolitical modes of borders as tools, networks, and worldviews. As infrastructures, borders are more than dispersed, variegated, and proliferated entities; they are movable entities traveling between places and spaces, between centers and peripheries, inside and outside Europe, and between visibility and invisibility. While border infrastructures detect traffic, they also disguise this traffic by highlighting some movements, but not others. As tools, borders are used in response to COVID-19 to enforce particular behaviors and to restructure human mobility, as instruments in international politics and diplomacy, and as bargaining chips to negotiate the possibility of mobility. Border infrastructures thus merge with a variety of other networks and data, information, and knowledge infrastructures to create interoperable dashboards. Once again, interoperability seems to be less concerned about arriving at an omniscient overview of situations than about facilitating specific interventions.

Historians have been quick to point out that border controls tend to outlast whatever crisis they are supposed to prevent. From the responses of cities and city-states during the plague to the Habsburg monarchy's *cordon sanitaire* along its frontier with the Ottoman Empire, history contains numerous examples of cordons, closures, isolations, and quarantines.[11] History is indeed replete with instances of states declaring exceptional measures in emergency situations that continue long after circumstances reach some form of normality. While the COVID-19 pandemic may foreshadow longer-term restrictions on international mobility and a political opportunity to enforce stricter migration policies, the idea of "extreme infrastructure in intra-ordinary situations" suggests that what has been initiated in response to COVID-19 may also be a harbinger of our future. Future conflicts, pandemics, climate change, and environmental disasters will lead to novel forms of international mobility, not only of humans but of other species, viruses, bacteria, and parasites. Borders as infrastructure will likely expand their technopolitical repertoires to intervene in more and more situations involving other forms of hybrid mobility.

Notes

Preface

1. The description of the COVID-19 pandemic is dated at June 28, 2020.

2. For an overview of Wildschut's photographic work and his projects on borders, shelter, and the Ville de Calais, see his website, accessed June 26, 2020, http://www.henkwildschut.com/.

3. "The Plants That Make Refugee Camps Feel More Like Home," *The New Yorker*, January 28, 2020, accessed June 26, 2020, https://www.newyorker.com/culture/photo-booth/the-plants-that-make-refugee-camps-feel-more-like-home.

4. "Biography Henk Wildschut," accessed June 26, 2020, http://www.henkwildschut.com/work/biography/biography-henk-wldschut/.

Chapter 1

1. "300 March across the Italian Border at Montegenevre," Calais Migrant Solidarity, April 23, 2018, https://calaismigrantsolidarity.wordpress.com/2018/04/23/300-march-across-the-italian-border-at-montgenevre/. Following the protest, several activists were arrested for helping foreigners enter French territory as part of an organized gang. "The Trial of Solidarity at the Italian-French Border in Southern Alps," Comité de soutien 3 + 4, accessed May 26, 2020, https://www.relaxepourles3plus4.fr/english/ For pictures of previous marches by Antonio Masiello, see "Alpine Crossing: Refugees Battle Extreme Weather to Reach France," *The Guardian*, January 26 2018, https://www.theguardian.com/world/gallery/2018/jan/26/alpine-crossing-refugees-battle-extreme-weather-reach-france.

2. The website of Generation Identitaire has videos of two patrolling helicopters and sport utility vehicles (SUVs) with the slogan "Defend Europe: Mission Alpes," and dozens of volunteers building a provisionary fence, a symbolic border in the snowy mountains. As the organization explains on its website, "first we closed the border to prevent illegal immigrants from entering France. Then we deployed in mobile

surveillance teams to monitor a larger area. Finally, we conducted investigations in order to understand where the illegal immigrants were passing by, who was smuggling them and how all this was organised in order to denounce them." Generation Identitaire, accessed May 26, 2020, https://generationidentitaire.org/.

3. It was not the first time that Generation Identitaire organized a mission. In March 2016, the group barricaded bridges to stop migrants entering Calais. In the summer of 2017, its members sought to prevent nongovernmental organization (NGO) ships from aiding migrants in the Mediterranean. Generation Identitaire, accessed May 26, 2020, https://generationidentitaire.org/.

4. "French Court Jails Far-Right Activists over Anti-immigrant Alps Stunt," *The Guardian*, August 26, 2019, https://www.theguardian.com/world/2019/aug/29/french-court-jails-far-right-activists-over-anti-migrant-alps-stunt.

5. Médecins Sans Frontières, *Harmful Borders: An Analysis of the Daily Struggle of Migrants as They Attempt to Leave Ventimiglia for Northern Europe* (Milan/Rome: MSF, 2018).

6. "Borders," Henk Wildschut, accessed May 26, 2020, http://www.henkwildschut.com/work/borders-7/.

7. "Alpine Migrant Route into France a Dead-End for Many," *The New Humanitarian*, November 30, 2016, https://www.thenewhumanitarian.org/feature/2016/11/30/alpine-migrant-route-france-dead-end-many.

8. "Migrants: La Roya, Vallée Rebelle," *Libération*, November 22, 2016, https://www.liberation.fr/france/2016/11/22/migrants-la-roya-vallee-rebelle_1529973?xtor=rss-450. A famous "rebel" is French olive and poultry farmer Cédric Herrou. His farm is located in Breil-sur-Roya, on the border of Italy and France. Part of a network of villagers, Herrou has been helping migrants make their way north. Their actions of civil disobedience fit into a tradition of trespassing into privileged spaces that ought to be publicly admissible. Natasha Basu, *Is This Civil? Transnationalism, Migration and Feminism in Civil Disobedience* (PhD thesis, University of Amsterdam, 2019), 75–76.

9. "Police officers, and even the foreign legion, have been deployed in and around the Roya Valley. Officially, they are fighting terrorism as part of the government's Operation Sentinel [an operation that stationed 10,000 troops on national territory]. In practice, soldiers patrol train stations and small roads, while police officers check trains and set up roadblocks, effectively sealing all exits out of the valley." "Alpine Migrant Route into France a Dead-End for Many," IRIN, November 30, 2016, https://www.refworld.org/docid/58403d9d4.html.

10. Human Rights Watch, *Subject to Whim: The Treatment of Unaccompanied Children in the French Hautes-Alpes* (New York: HRW, 2019), 1–4.

11. "France: Migrants at the Frozen Border," *BBC*, December 12, 2017, https://www.bbc.com/news/av/world-europe-42317793/france-migrants-at-the-frozen-border.

12. My use of the notion of "vehicles" shares certain similarities with the notion of "interscalar vehicles" introduced by Gabrielle Hecht in "Interscalar Vehicles for an African Anthropocene: On Waste, Temporality, and Violence," *Cultural Anthropology* 33, no. 1 (2018): 135. According to Hecht, interscalar vehicles are "objects and modes of analysis that permit scholars and their subjects to move simultaneously through deep time and human time, through geological space and political space."

13. Robert MacFarlane, *Mountains of the Mind: Adventures in Reaching the Summit* (New York: Vintage Books, 2004), 144.

14. "Europe's borders, like all borders, are the materializations of sociopolitical relations that mediate the continuous production of the distinction between the putative inside and outside, and likewise mediate the diverse mobilities that are orchestrated and regimented through the production of that spatial divide." Nicholas De Genova, "Introduction. The Borders of 'Europe' and the European Question," *The Borders of "Europe": Autonomy of Migration, Tactics of Bordering*, ed. Nicholas De Genova (Durham/ London: Duke University Press, 2017), 21.

15. Isabelle Stengers, *The Invention of Modern Science* (Minneapolis/London: University of Minnesota Press, 2000), 99–100.

16. Donna Haraway, "Situated Knowledges: The Science Question in Feminism and the Privilege of Partial Perspective," *Feminist Studies* 14, no. 3 (1988): 575–599; Bruno Latour, "On Technical Mediation," *Common Knowledge* 3, no. 2 (1994): 29–64; Annemarie Mol, "Actor-Network Theory: Sensitive Terms and Enduring Tensions," *Kölner Zeitschrift für Soziologie und Sozialpsychologie* 50, no. 1 (2010): 253–269; Peter-Paul Verbeek, *What Things Do: Philosophical Reflections on Technology, Agency, and Design* (University Park: Pennsylvania State University Press, 2005). For a discussion of the notion of mediation, see Yoni van den Eede, "In between Us: On the Transparency and Opacity of Technological Mediation," *Foundations of Science* 16, no. 2–3 (2011): 139–159.

17. In Mexican immigrants' rights movements and in Chicano literature and popular culture, there is a saying that "We didn't cross the border, the border crossed us." The aphorism suggests that borders can cut through existing communities and territories, whether through natural or material means (a wall, fence, or river), administrative fiat (permits or rights to services such as schools and healthcare), economic sanctions (taxes or import duties), or sociocultural boundaries (gender, language, or race). The aphorism also underlines that the study of borders has everything to do with movement and mobility, of migrants and borders alike. See, for instance, Roberto Delgadillo Hernández, *Coloniality and Border(ed) Violence: San Diego, San Ysidro and the U-S/// Mexico Border* (Tucson: The University of Arizona Press, 2018); University of California at Berkeley, 2010), and Josue David Cisneros, *Rhetorics of Borders, Citizenship and Latina/o Identity* (Tuscaloosa: University of Alabama Press, 2013).

18. Thomas Nail, *Theory of the Border* (New York: Oxford University Press, 2016).

19. Nail, *Theory of the Border*, 3–7.

20. Matthew Longo, *Politics of Borders. Sovereignty, Security, and the Citizen after 9/11* (Cambridge: Cambridge University Press, 2018), 69–70.

21. "The border," Longo argues, "is generative of political space; it is also the site of politics—of violence, of technologies of control and the architecture of the state." Longo, *Politics of Borders*, 43.

22. "Viapolitics is not a synonym for the biopolitics of mobility, migration politics, or the autonomy of migration, though it does intersect these domains. What it does represent is a particular angle of inquiry, one that treats the interaction of humans and vehicles as an irreducible feature of migratory struggles. At the same time it is not a call for a general theory about migration, vehicles and politics. It is not a general theory because the ways in which vehicles, struggles and power interact is quite heterogeneous and defies any simplistic schema." William Walters, "Migration, Vehicles, and Politics: Three Theses on Viapolitics," *European Journal of Social Theory* 18, no. 4 (2015): 15.

23. As Walters points out, "a vehicle can become a mobile border zone" while border zones themselves are often movable entities. "New objects and understandings come into view once we see migration from the angle of its routes and vehicles." Walters, "Migration, Vehicles, and Politics."

24. Hein de Haas, Katharina Natter, and Simona Vezzoli, "Growing Restrictiveness or Changing Selection? The Nature and Evolution of Migration Policies," *International Migration Review* 52, no. 2 (June 2018): 324–367.

25. "Norway Tells Refugees Who Used Cycling Loophole to Enter to Return to Russia," *The Guardian*, January 14, 2016, https://www.theguardian.com/world/2016/jan/14/norway-tells-refugees-bikes-russia-bicycle-immigration-storskog.

26. "Regimes of movement are thus never simply a way to control, to regulate, or to incite movement. Regimes of movement are integral to the *formation of different modes of being*." As such, movement "allows different bodies to take form." Hagar Kotef, *Movement and the Ordering of Freedom: On Liberal Governances of Mobility* (Durham, NC: Duke University Press, 2015), 14–15 (italics in the original).

27. Movements do not only entail the traveling of people. Movements also include the composition of bodies and assemblages. Composition indicates that "things" that bring together humans and technologies must be mediated so as to become part of a constellation. As Kotef suggests, "Standing as an opposition to nature, to stable power structures, to static state bureaucracy, politics brings the potential carried by instability: the potential of change, of widening the gaps allowing our agency, redistributing resources, and realigning power. A set of different (even if tangent) traditions of thinking about the meaning of 'the political' conceptualizes it as that which moves, as the moment of movement, or as that to which movement is essential."

Kotef, *Movement and the Ordering of Freedom*, 13. In my view, the relationships between politics and nature, and between humans and technology, are less dichotomous than Kotef suggests.

28. Simon Cole, *Suspect Identities: A History of Fingerprinting* (Cambridge, MA: Harvard University Press, 2001), 63–70. The United States began fingerprinting foreigners including Chinese immigrants in the nineteenth century before applying it to other populations for other purposes. See Cole, *Suspect Identities*, 119–139.

29. For a telling account about "waiting" at borders, see Shahram Khosravi, "Waiting: Keeping Time," *Migration: A COMPAS Anthology*, ed. B. Anderson and M. Keith (Oxford, UK: COMPAS, 2014).

30. Katy Hayward, "A Frictionless Border Is Impossible, an Invisible One Undesirable," *Irish Times*, July 15, 2017.

31. The distinction originates from Martin Heidegger, "The Question Concerning Technology," *The Question Concerning Technology and Other Essays* (New York: Harper Perennial, 1977), 3–35. It is not without hesitation that I refer to Heidegger. It may seem unfortunate, to say the least, to start a discussion of technopolitics and borders by rereading his work. It is clear that Heidegger, who is widely considered one of the most important thinkers of the twentieth century, not only collaborated with the Nazi regime, but also sympathized with it intellectually and ideologically. The publication of his *Schwarze Hefte* ("black notebooks") revealed much of the German philosopher's personal thoughts during the defining period of 1931 until 1941. The black notebooks added a deeper and unquestionably more intimate chapter to the philosopher's already damaged reputation and identified his philosophy as contaminated with Nazism, a blemish from which he never cleaned himself. Nonetheless, Heidegger's thinking has been extremely influential and his emphasis on ontology has helped us reinterpret the conceptual history of Western philosophy and his "ontological distinction" remains an important, albeit debatable, contribution.

32. Mark Brown, "Politicizing Science: Conceptions of Politics in Science and Technology Studies," *Social Studies of Science* 45, no. 1 (2015): 9.

33. Marieke de Goede's "The Politics of Preemption and the War on Terror in Europe," *European Journal of International Relations* 14, no. 1 (2008): 161–85 reminds us of the meaning of preemption in the war on terror and in European counterterror policies. Nick Vaughan-Williams's *Border Politics: The Limits of Sovereign Power* (Edinburgh: Edinburgh University Press, 2009) relates border politics to Foucault's understandings of security, territory and law, biopolitics and geopolitics, violence, and sovereign power. The "politics of possibility" in Louise Amoore's *The Politics of Possibility: Risk and Uncertainty beyond Probability* (Durham, NC: Duke University Press, 2013) alerts us to the security situations, books, and artworks concerned with unknown futures that arose after 9/11. The "politics of techniques, devices and acts" expressed in Claudia Aradau and Jef Huysmans's "Critical Methods in International

Relations: The Politics of Tools, Devices and Acts," *European Journal of International Relations* 20, no. 3 (2014): 596–619 critically examines the conceptual repertoire of international relations and security studies, and considers methods themselves as devices and acts. The "post/humanitarian border politics" in Vicki Squire's *Post/ Humanitarian Border Politics between Mexico and the US: People, Places, Things* (New York: Palgrave Macmillan, 2015) invites us to visit the US-Mexico border region and the people, places, and things that affect the materiality, territory, and politics of the borderlands. The "politics of counting" in Martina Tazzioli's "The Politics of Counting and the Scene of Rescue. Border Deaths in the Mediterranean," *Radical Philosophy* 192 (June/August 2015) discusses various border death statistics and the gap between countable and uncountable deaths. The "politics of security technology" in Ruben Andersson's "Hardwiring the Frontier? The Politics of Security Technology in Europe's 'Fight against Illegal Migration,'" *Security Dialogue* 47, no. 1 (2016): 22–39 analyzes the security technology of the "Seahorse network" that "hardwires" border cooperation into a satellite system connecting African and European forces, and shows how the initiative multiplies the involvement of actors and creates new forms of collaboration, competition and conflict. The "politics of prediction" in Claudia Aradau and Tobias Blanke's "Politics of Prediction: Security and the Time/Space of Governmentality in the Age of Big Data," *European Journal of Social Theory* 20, no. 3 (2016): 373–91 leads us to the field of security, Big Data, and the digital mediation of practices of policing, social control, and war. The "politics and the digital" in Mareile Kaufmann and Julien Jeandesboz's "Politics and 'the Digital': From Singularity to Specificity," *European Journal of Social Theory* 20, no. 3 (2016): 309–28 addresses mathematical functions, diagrams, and graphs, and examines the "relationship between 'the digital' and political acts" and the "new political subjects and subjectivities" this relationship brings about. The "kinopolitics" of Thomas Nail's *Theory of the Border* (New York: Oxford University Press, 2016) is a theory of motion that underlies the formation of societies and states and that distinguishes historically and systematically different types of borders at the US-Mexican border such as the fence, wall, cell and checkpoint. "Tempo politics," a concept introduced by Simon Sontowski's "Speed, Timing and Duration: Contested Temporalities, Techno-political Controversies and the Emergence of the EU's Smart Border," *Journal of Ethnic and Migration Studies* 44, no. 16 (2018): 2730–2746 concerns how the EU Smart Borders Package and the EU's bio databases problematize border control, primarily on the level of its temporalities.

34. For similar approaches to technopolitics and infrastructure, see Andrew Barry, *Material Politics: Disputes along the Pipeline* (Oxford, UK: Wiley-Blackwell, 2013); Timothy Mitchell, *Carbon Democracy: Political Power in the Age of Oil* (London: Verso, 2011); Gabrielle Hecht and Paul Edwards, "The Technopolitics of Cold War. Toward a Transregional Perspective," *Essays on Twentieth Century History* (Philadelphia: Temple University Press, 2010); and Gabrielle Hecht, *Entangled Geographies: Empire and Technopolitics in the Global Cold War* (Cambridge, MA: MIT Press, 2011). For Hecht, technopolitics is more than the "strategic practice of designing or using technology to enact political goals." It contains unpredictable dimensions that "exceed

or escape the intentions of system designers" as politics and technology morph into each other: "the material qualities of technopolitical systems shape the texture and the effects of their power." See Hecht, *Entangled Geographies*, 3.

35. Marilyn Strathern, *The Relation: Issues in Complexity and Scale* (Cambridge, UK: Prickly Pear Press, 1995), 27–28.

36. Mark Elam, "Living Dangerously with Bruno Latour in a Hybrid World," *Theory, Culture & Society* 16, no. 4 (1999): 1–24.

37. This notion of morphology has much in common with the views on technology of the German-American philosopher Ernst Kapp (1808–1896). Similar ideas were later advanced by Marshall McLuhan in his "extension theory" in Marshall McLuhan, *Understanding Media: The Extensions of Man* (Corte Madera: Gingko Press, 2003), and by Bruno Latour as well, but Kapp's work is much less known. A recent translation of his 1877 book *Elements of a Philosophy of Technology: On the Evolutionary History of Culture*, originally published in German, reveals Kapp's original and still inspiring views on the relationship between humans and technology. See Ernst Kapp, *Elements of a Philosophy of Technology: On the Evolutionary History of Culture* (Minneapolis: University of Minnesota Press, 2018).

38. Martin Heidegger, "Building, Dwelling, Thinking," *Basic Writings* (Abingsdon, UK: Routledge, 1993), 239–257.

39. Bruno Latour, *Reassembling the Social: An Introduction to Actor-Network-Theory* (Oxford: Oxford University Press, 2005).

40. One example is Ruben Andersson's analysis in *Illegality, Inc.: Clandestine Migration and the Business of Bordering Europe* (Oakland: University of California Press, 2014) of clandestine migration trails and the various political and technological associations that relate Europe's border regime to "illegal immigrants." Another is Marieke de Goede's "The Chain of Security," *Review of International Studies* 44, no. 1 (2018): 24–42, on detecting suspicious transactions to combat terrorism and other security treats. By following the actors and tracing the "chain of security," she brings the imperative to "follow the actors" back to the original slogan "follow the money."

41. See, for instance, Nicholas De Genova, "Introduction"; and Stephan Scheel, *Autonomy of Migration? Appropriating Mobility within Biometric Border Regimes* (London/New York: CRC Press, 2019).

42. Nicholas De Genova, "Introduction," 6.

43. Nicholas De Genova, "Introduction," 6.

44. Nicholas De Genova, "The Queer Politics of Migration: Reflections on 'Illegality' and Incorrigibility," *Studies in Social Justice* 4, no. 2 (2010): 101–126.

45. See "Geschiedenis van wij zijn hier," *Wij zijn hier*, accessed May 26, 2020, http://wijzijnhier.org/tijdslijn/geschiedenis-van-wij-zijn-hier/.

46. As Peter Nyers says, "The migrant is not the only mobile agent at the border. The border, too, moves." Peter Nyers, "Moving Borders: The Politics of Dirt," *Radical Philosophy* 174 (July/August 2012): 2.

47. Biao Xiang and Johan Lindquist, "Migration Infrastructure," *International Migration Review* 48, no. 1 (2014): 124, 132, maintain a symmetrical point of view: "it is not migrants who migrate, but rather constellations consisting of migrants and non-migrants, of human and non-human actors." They add: "Migration flows can be fragmented and short-lived, but infrastructure retains a particular stability and coherence."

48. According to Ruben Andersson in "Hardwiring the Frontier? The Politics of Security Technology in Europe's 'Fight against Illegal Migration,'" *Security Dialogue* 47, no. 1 (2016): 24–25, the framing of human and nonhuman groups as actants "allows for shifting the focus away from the two poles of migration studies—the (political science) view that privileges policy and the (ethnographic) insistence on a grounded 'migrants' perspective'—towards the material, virtual, and social interfaces of the migratory encounter. From this vantage point, the fences, control rooms, and data systems . . . act as mediators in a network."

49. Paul Edwards. *A Vast Machine: Computer Models, Climate Data, and the Politics of Global Warming* (Cambridge, MA: MIT Press, 2013), 84.

50. My analysis here corresponds with Sandro Mezzadra's and Brett Neilson's argument in *Border as Method, or the Multiplication of Labor* (Durham, NC: Duke University Press, 2013), 18, which states that the border is "not so much a research object as an epistemological viewpoint that allows an acute critical analysis not only of how relations of domination, dispossession, and exploitation are being redefined presently but also of the struggles that take shape around those changing relationships."

51. Mezzadra and Neilson, *Border as Method*.

52. See Mathias Czaika and Hein de Haas, "The Globalization of Migration: Has the World Become More Migratory?" *International Migration Review* 48 (2014): 283–323.

53. "The 10-Point Action Plan in Action, 2016–Glossary," United Nations High Commissioner for Refugees, December 2016, 282, accessed May 28, 2020, https://www .unhcr.org/asylum-and-migration.html. See also "Glossary of Migration," *International Organization of Migration* (Geneva, Switzerland: IOM, 2019), 141–142.

54. "Asylum and Migration Glossary 6.0," European Commission, May 2018, 263, https://ec.europa.eu/home-affairs/what-we-do/networks/european_migration_network/glossary_en.

55. As Anthony Cooper, Chris Perkins, and Chris Rumford point out in "The Vernacularization of Borders," *Placing the Border in Everyday Life*, ed. Reece Jones and Corey Johnson (New York: Routledge, 2016), 18, "The EU has emerged as a major actor in the business of creating, relocating, and dismissing borders. The EU shifts the borders

of Europe every time it enlarges, it turns national borders into European borders, it regulates and harmonizes European borders through Frontex, its border agency, and it has the power to decide where the important borders in Europe are to be found."

56. For a discussion of the relationships that states and empires have with borders, see Charles Maier, *Once within Borders: Territories of Power, Wealth and Belonging Since 1500* (Cambridge, MA: Harvard University Press, 2016).

57. Thomas Gammeltoft-Hansen and Nina Nyberg Sørensen describe this in *The Migration Industry and the Commercialization of International Migration* (New York: Routledge, 2013) as the rise of the "migration industry." Marijn Hoijtink, in *Securing the European "Homeland": Profit, Risk, Authority* (PhD thesis, University of Amsterdam, 2016), sees the privatization and commercialization of all kinds of tasks and technologies as the emergence of a "homeland security market." Gallya Lahav and Vriginie Guiraudon, "Comparative Perspectives on Border Control: Away from the Border and Outside the State," in *The Wall around the West. State Borders and Immigration Controls in North America and Europe*, ed. Peter Andreas and Timothy Snyder (Lanham: Rowman and Littlefield Publishers, 2000); and "Actors and Venues in Immigration Control: Closing the Gap between Political Demands and Policy Outcomes," *West European Politics* 29, no. 2 (2006): 201–223, detail how the governance of migration is shifting to nonstate actors, such as to airlines, transport companies, and employers. Biao Xiang and Johan Lindquist, writing about migration infrastructures in Indonesia and China in "Migration Infrastructure," *International Migration Review* 48, no. 1 (2014): 137, view the delegation of responsibility to local intermediaries as a strategy that turns companies into functioning arms of the state.

58. See, for instance, Amade M'charek, Katharina Schramm, and David Skinner, "Topologies of Race: Doing Territory, Population and Identity in Europe," *Science, Technology, & Human Values* 39, no. 4 (2014): 468–487.

Chapter 2

1. See Dennis Broeders, *Breaking Down Anonymity. Digital Surveillance of Irregular Migrants in Germany and the Netherlands* (Amsterdam: Amsterdam University Press, 2009) for a discussion of these concepts.

2. This is presumably what Frank Schipper and Johan Schot had in mind by using the term "infrastructural Europeanism." See "Infrastructural Europeanism, or the Project of Building Europe on Infrastructures: An Introduction," *History and Technology* 27, no. 3 (2011): 245–264.

3. See, for instance, Michael Thad Allen and Gabrielle Hecht, *Technologies of Power. Essays in Honor of Thomas Parke Hughes and Agatha Chipley Hughes* (Cambridge, MA: MIT Press, 2001), and Erik van der Vleuten and Arne Kaijser, "Networking Europe," *History and Technology* 21, no. 1 (2005): 21–48.

4. See Irene Anastasiadou, *Constructing Iron Europe: Transnationalism and Railways in the Interbellum* (Amsterdam: Amsterdam University Press, 2011); Jiří Janác, *European Coasts of Bohemia: Negotiating the Danube-Oder-Elbe Canal in a Troubled Twentieth Century* (Amsterdam: Amsterdam University Press, 2012); Vincent Lagendijk, *Electrifying Europe: The Power of Europe in the Construction of Electricity Networks* (Amsterdam: Amsterdam University Press, 2008); Suzanne Lommers, *Europe on Air: Interwar Projects for Radio Broadcasting* (Amsterdam: Amsterdam University Press, 2012); and Judith Schueler, *Materialising Identity: The Co-construction of the Gotthard Railway and Swiss National Identity* (Eindhoven, Netherlands: Technische Universiteit Eindhoven, 2008).

5. I refer here to the research network "Tensions of Europe," accessed May 27, 2020, https://www.tensionsofeurope.eu/.

6. According to Paul Edwards in "Infrastructure and Modernity: Scales of Force, Time, and Social Organization in the History of Sociotechnical Systems," in *Modernity and Technology*, ed. Thomas Misa, Philip Brey, and Andrew Feenberg (Cambridge MA: MIT Press, 2003), 186: "by linking macro, meso, and micro scales of time, space, and social organization . . . infrastructures simultaneously shape and are shaped by—in other words, co-construct—the condition of modernity."

7. In "Infrastructure and the State in Science and Technology Studies," *Social Studies of Science* 45, no. 1 (2015): 137–145, Nicholas Rowland and Jan-Hendrik Passoth point to Sara Pritchard's *Confluence: The Nature of Technology and the Remaking of the Rhône* (Cambridge, MA: Harvard University Press, 2011); Jo Guldi's *Roads to Power: Britain Invents the Infrastructure State* (Cambridge, MA: Harvard University Press, 2012); Allan Mazur's *Energy and Electricity in Industrial Nations: The Sociology and Technology of Energy* (London: Routledge, 2013); and Andrew Barry's *Material Politics: Disputes along the Pipeline* (Oxford, UK: Wiley-Blackwell, 2013).

8. See Edwards, "Infrastructure and Modernity," 12.

9. As Polly Pallister-Wilkins argues in "How Walls Do Work: Security Barriers as Devices of Interruption and Data Capture," *Security Dialogue* 47, no. 2 (2016): 154: "barriers are not just barriers, fences or walls: they include openings, checkpoints and gates that allow for the movement of people and goods. This allowance and the subsequent governance of the comings and goings of people and goods suggests that barriers are not simply concerned with prescribing, securing and administering the *intra muros* . . . but with governing movement across them."

10. As Brian Larkin notes in "The Politics and Poetics of Infrastructure," *Annual Review of Anthropology* 42 (2013): 327, 330: "infrastructures are material forms that allow for the possibility of exchange over space," and "typically begin as a series of small, independent technologies with widely varying technical standards."

11. As Didier Bigo argues in "Death in the Mediterranean Sea: The Results of the Three Fields of Action of European Union Border Controls," in *The Irregularization of*

Migration in Contemporary Europe, ed. Jansen Yolande, Robin Celikates, and Joost de Bloois (London: Rowman and Littlefield, 2015), 59–60, borders can be "solid," "liquid," or "cloudy." These do not necessarily coincide with where border control takes place (land, sea, or air), but distinguish among three "fields of action." The first is the conception of the border as a solid barrier, related to an idea of "defense." The second concerns border checks and practices of "policing and surveillance," including processes of identifying, authenticating, and filtering. The third is "the universe of the transnational database." Connected "to the digital and the virtual, to data doubles and their cohorts, to categorizations resulting from algorithms, to anticipations of unknown behaviors, to the prevention of future actions," it "pervades the second universe (and sometimes the first) by justifying technology and the management of surveillance at a distance in the name of the protection of a group of the population." From this point of view, "borders are pixels. The sea no longer exists."

12. Addressing Bruno Latour's theory of associations and his political theory of collectives, Sven Opitz and Uwe Tellman, "Europe as Infrastructure: Networking the Operative Community," *South Atlantic Quarterly* 141, no. 1 (2015): 186–187, aim to bring "communitarian discourses to the level of technological works" in order to "turn these technical matters of infrastructural politics into proper matters of concern," as Latour calls them. The notions of "connectivity" and "collectivity" speak to Latour's program of making things public, and thus to a visual approach to *Dingpolitik*.

13. Hein De Haas, Katharina Natter, and Simona Vezzoli, "Growing Restrictiveness or Changing Selection? The Nature and Evolution of Migration Policies," *International Migration Review* 52, no. 2 (June 2018): 324–367.

14. Larkin ["The Politics and Poetics of Infrastructure," 336] suggests an aesthetic, anti-dualist maneuver to transcend the visible-invisible distinction in the working of technologies: "invisibility is certainly one aspect of infrastructure, but it is only one and at the extreme edge of a range of visibilities that move from unseen to grand spectacles and everything in between."

15. As Olga Kuchinskaya argues in *The Politics of Invisibility: Public Knowledge about Radiation Health Effects after Chernobyl* (Cambridge, MA: MIT Press, 2014), a certain "politics of invisibility" informs the making and unmaking of public events and the shaping of public knowledge. This interplay between visibility and invisibility in infrastructure theory is close to Jacques Rancière's notion of the "distribution of the sensible," as developed in *Dissensus: On Politics and Aesthetics* (London: Bloomsbury, 2010), 12, 70, 139. Based on the interplay between the visible and the invisible—on what is represented and what remains disclosed—politics comes into being once a given order of things is contested. Politics "re-frames the given by inventing new ways of making sense of the sensible, new configurations between the visible and the invisible, and between the audible and the inaudible, new distributions of space and time." What becomes visible in such configurations is what Rancière calls "the part that has no part." A similar circumscription of politics occurs in international border control and mobility policies.

16. Biao Xiang and Johan Lindquist, "Migration Infrastructure," *International Migration Review* 48, no. 1 (2014): 124.

17. This framing of humans and nonhumans as actants "allows for shifting the focus away from the two poles of migration studies—the (political science) view that privileges policy and the (ethnographic) insistence on a grounded 'migrants' perspective'— towards the material, virtual, and social interfaces of the migratory encounter. From this vantage point, the fences, control rooms, and data systems . . . act as mediators in a network," argues Ruben Andersson in "Hardwiring the Frontier? The Politics of Security Technology in Europe's 'Fight against Illegal Migration,'" *Security Dialogue* 47, no. 1 (2016): 24–25.

18. "Externalization" is analogous to "remote control" in Aristide Zolberg's "The Archaeology of 'Remote Control,'" in *Migration Control in the North Atlantic World: The Evolution of State Practices in Europe and the United States from the French Revolution to the Inter-War Period*, ed. Andreas Fahrmeier, Oliver Faron, and Patrick Weil (New York: Berghahn Books, 2003) and "policing at a distance," as discussed by Elspeth Guild and Didier Bigo in "The Transformation of European Border Controls," in *Extraterritorial Immigration Control. Legal Challenges*, ed. Bernard Ryan and Valsamis Mitsilegas (Leiden, Netherlands: Brill Academic Publishers, 2010), 258.

19. See for instance, "Asylum and Migration in the EU: Facts and Figures," *European Parliament*, June 30, 2016 (updated July 22, 2019), http://www.europarl.europaeu /news/en/headlines/society/20170629STO78630/eu-migrant-crisis-facts-and-figures; and "Mediterranean the Deadliest Sea for Refugees and Migrants, says UN Agency," *United Nations*, January 31, 2012, https://news.un.org/en/story/2012/01/401822 -mediterranean-deadliest-sea-refugees-and-migrants-says-un-agency.

20. "Schengen Museum Ceiling Collapses—but It's Not a Sign, Mayor Says," *The Guardian*, May 16, 2016, https://www.theguardian.com/world/2016/may/16/schengen -museum-ceiling-collapse-bad-omen-europe-borders.

21. While migration was of secondary concern, the raising of the Iron Curtain was likely to stimulate migration from Eastern Europe. Western European governments feared a potential so-called invasion from the East. For that reason, EU member-states began searching for durable solutions. Various member-states were also concerned about migrant workers and the legal status of guest workers and their families. The European Court of Justice had been growing increasingly assertive on the issue of the free movement of European Community citizens, ruling to restrict existing provisions by which member-states could expel people on the grounds of public policy, security, and health. See Ruben Zaiotti, *Cultures of Border Control. Schengen and the Evolution of European Frontiers* (Chicago: University of Chicago Press, 2011), 62.

22. Italy had to wait to join the Schengen Area until 1997, and Greece until 2000, as provisions had to be made at their external borders. The Scandinavian countries joined in 2011. Of the twenty-six countries that are now part of the Schengen Area,

twenty-two have fully implemented the Schengen acquis. Iceland, Norway, Switzerland, and Lichtenstein are members of the European Free Trade Association (EFTA) and associate members of the Schengen Area, but they are not members of the European Union. They implement the Schengen acquis through specific agreements.

23. See "The Schengen acquis - Convention implementing the Schengen Agreement of 14 June 1985 between the Governments of the States of the Benelux Economic Union, the Federal Republic of Germany and the French Republic on the gradual abolition of checks at their common borders," *Official Journal L 239 , 22/09/2000 P. 0019 – 0062*, accessed November 5, 2020, https://eur-lex.europa.eu/LexUriServ/LexUri Serv.do?uri=CELEX%3A42000A0922%2802%29%3AEN%3AHTML.

24. As Guild and Bigo point out in "The Transformation of European Border Controls," 266, "the definition of the external border is exclusively by reference to the definition of the internal border."

25. See Zaiotti, *Cultures of Border Control*, 71–72.

26. According to Eurostat, between 2008 and 2012, there was a gradual increase in the number of asylum applications within the EU-27, after which the number of asylum seekers rose at a more rapid pace, with 400,500 applications in 2013, 594,200 in 2014, and then around 1.3 million in 2015 and 1.2 million in 2016. In 2017, the number of asylum applications saw a decrease of 44.5 percent in comparison with 2016, and continued a downward path in 2018. See "Asylum Applications (Non-EU) in the EU-27 Member States, 2008–2019," *Eurostat*, accessed May 27, 2020, https://ec .europa.eu/eurostat/statistics-explained/index.php/Asylum_statistics.

27. Ernst Hirsch Ballin, Emina Ćerimović, Huub Dijstelbloem, and Mathieu Segers, *European Variations as a Key to Cooperation* (Cham, Switzerland: Springer Open, 2020), 146.

28. "Technologies always embody compromise," Wiebe Bijker and John Law argue in *Shaping Technology/Building Society: Studies in Sociotechnical Change* (Cambridge, MA: MIT Press, 1992), 3, when describing the simultaneous development of technologies and societies.

29. The concept of imagination in this case shares some similarities with the idea of "sociotechnical imaginaries." Sheila Jasanoff and Sang-Hyun Kim, in "Containing the Atom: Sociotechnical Imaginaries and Nuclear Power in the United States and South Korea," *Minerva* 47, no. 2 (2009): 120, defined sociotechnical imaginaries in a study of US and South Korean responses to nuclear power as "collectively imagined forms of social life and social order reflected in the design and fulfillment of nation-specific scientific and/or technological projects." Whereas this notion expresses the entanglement between the social and technological order, it also maintains an emphasis on national projects that unites these orders in specific ways. However, imaginaries/imaginations do not emerge only from national configurations. In a

later study, "Future Imperfect," in *Dreamscapes of Modernity*, ed. Sheila Jasanoff and Sang Hyun Kim (Chicago: University of Chicago Press, 2015), 4, Sheila Jasanoff redefined sociotechnical imaginaries as "collectively held, institutionally stabilized, and publicly performed visions of desirable futures, animated by shared understandings of forms of social life and social order attainable through, and supportive of, advances in science and technology."

30. The movement of political ideas through all kinds of practices, symbols, and materialities was vividly described by Benedict Anderson in his classic study of nationalism, *Imagined Communities* (New York: Verso, 1983). Political ideas about nations and nationality proved to be highly mobile, firing the imaginations of people around the world through words, meanings, symbols, and technologies. The relationship between the form and content of politics has been addressed by Arjun Appadurai. In *The Future as Cultural Fact. Essays on the Global Condition* (New York: Verso, 2013), 64, he argues that it is crucial to understand the relationship "between the forms of circulation and the circulation of forms." In his discussion of "moving geographies," he points out that actors and things gain political significance as they travel, and that humans, words, images, and ideas often travel using different routes and vehicles. In the two volumes edited by Mark Salter, *Making Things International 1: Circuits and Motion* and *Making Things International 2: Catalysts and Reactions* (Minneapolis/London: University of Minnesota Press, 2015), the interaction between forms and circuits is studied through the analysis of all kinds of devices and techniques, ranging from passport photos and hotlines to drones and barbed wire.

31. Inspiration can be found in literature that combines the study of material geopolitical and governance issues with a science and technology approach. In *Political Machines: Governing a Technological Society* (London: Athlone Press, 2001), 67, Andrew Barry argues that "the difficulty in understanding the European economy and the European political system has been, at least in part, a function of the critical part played by a vast array of objects and technical devices in its make-up." In *Carbon Democracy: Political Power in the Age of Oil* (London/New York: Verso, 2011), Timothy Mitchell unpacks the substance of oil as well as the idea of democracy in a symmetrical move, demonstrating that both rely on specific practices in which the composing elements are processed and gathered together.

32. See "Mission and Tasks," Eurojust, accessed May 26, 2020, http://www.eurojust .europa.eu/about/background/Pages/mission-tasks.aspx.

33. The seven ad hoc centers were Risk Analysis Centre (Helsinki), Centre for Land Borders (Berlin), Air Borders Centre (Rome), Western Sea Borders Centre (Madrid), Ad-hoc Training Centre for Training (Traiskirchen, Austria), Centre of Excellence (Dover, United Kingdom), and Eastern Sea Borders Centre (Piraeus, Greece). See "Origin," Frontex, accessed January 15, 2018, http://frontex.europa.eu/about-frontex /origin/.

34. Gregory Feldman, *The Migration Apparatus* (Stanford, CA: Stanford University Press, 2012), 84.

35. "Partners & Agreements: Frontex," Europol, accessed June 1, 2020, https://www.europol.europa.eu/agreements/frontex.

36. Maarten den Heijer, "Frontex and the Shifting Approaches to Boat Migration in the European Union: A Legal Analysis," *Externalizing Migration Management. Europe, North America and the Spread of "Remote Control" Practices*, ed. Ruben Zaiotti (London/New York: Routledge, 2016), 53.

37. den Heijer, "Frontex and the Shifting Approaches," 67.

38. Charles Heller, Lorenzo Pezzani, and Situ Studio, "Report on the 'Left-to-Die-Boat,'" *Forensic Architecture Project Goldsmiths University of London*, April 11, 2012; Tineke Strik, "Parliamentary Assembly of the Council of Europe"; *Lives Lost in the Mediterranean Sea: Who Is Responsible?* Doc. 12895. April 5, 2012; den Heijer, "Frontex and the Shifting Approaches"; David Scott FitzGerald, *Refuge beyond Reach. How Rich Democracies Repel Asylum Seekers* (Oxford: Oxford University Press, 2019).

39. The event fits in with what FitzGerald, *Refuge beyond Reach*, 6, calls the European Union's "architectures of repulsion."

40. Heller et al., "Report on the 'Left-to-Die-Boat.'"

41. den Heijer, "Frontex and the Shifting Approaches," 58–60.

42. Strik, "Parliamentary Assembly of the Council of Europe."

43. "Wreck of Migrant Ship That Killed Hundreds Will Be Displayed at Venice Biennale," *The New York Times*, May 6, 2019, https://www.nytimes.com/2019/05/06/arts/design/migrant-boat-venice-biennale-christian-buchel.html.

44. "Joint Foreign and Home Affairs Council: Ten Point Action Plan on Migration," *European Commission*, press release, April 20, 2015, http://europa.eu/rapid/press-release_IP-15-4813_en.htm.

45. Michela Ceccorulli and Sonia Lucarelli, "Securing the EU's Borders in the Twenty-First Century," in *EU Security Strategies: Extending the EU System of Security Governance*, ed. Spyros Economides and James Sperling (Abingdon, UK: Routledge, 2018).

46. Henriette Ruhrmann and David FitzGerald, "The Externalization of Europe's Borders in the Refugee Crisis, 2015–2016," San Diego, CA, California Univeristy Center for Comparative Immigration Studies, 2017, 26–28.

47. Mariana Gkliatia and Herbert Rosenfeldt, "Accountability of the European Border and Coast Guard Agency: Recent Developments, Legal Standards and Existing Mechanisms," *School of Advanced Study University of London. RLI Working Paper No. 30*, 2018, 13.

48. REGULATION (EU) 2019/1896 OF THE EUROPEAN PARLIAMENT AND OF THE COUNCIL of 13 November 2019 on the European Border and Coast Guard, and repealing Regulations (EU) No 1052/2013 and (EU) 2016/1624.

49. "Schengen Information System," *European Commission*, accessed May 27, 2020, https://ec.europa.eu/home-affairs/what-we-do/policies/borders-and-visas/schengen -information-system_en.

50. Dennis Broeders, "The New Digital Borders of Europe: EU Databases and the Surveillance of Irregular Migrants," *International Sociology* 22, no. 1 (2007): 7.

51. Broeders, "The New Digital Borders of Europe," 9.

52. "Schengen Information System."

53. "Eurodac," *European Data Protection Supervisor*, accessed May 27, 2020, https://edps.europa.eu/data-protection/data-protection/glossary/e_en.

54. Dennis Broeders and Huub Dijstelbloem, "The Datafication of Mobility and Migration Management: The Mediating State and Its Consequences," in *Digitizing Identities: Doing Identity in a Networked World*, ed. Irma van der Ploeg and Jason Pridmore (New York/London: Routledge, 2015), 242–260.

55. Georgios Glouftsios, "Designing Digital Borders: The Visa Information System (VIS)," in *Technology and Agency in International Relations*, edited by Marijn Hoijtink and Matthias Seele (London/New York: Routledge, 2019), 178–179.

56. Broeders and Dijstelbloem, "The Datafication of Mobility and Migration Management."

57. See Huub Dijstelbloem and Albert Meijer (Eds.), *Migration and the New Technological Borders of Europe* (Basingstoke, UK: Palgrave Macmillan, 2011); and Michiel Besters and Frans Brom, "'Greedy' Information Technology: The Digitalization of the European Migration Policy," *European Journal of Migration and Law* 12, no. 4 (2010): 455–470.

58. According to the proposal, "the EES should apply to third-country nationals, both visa-required and visa-exempt travelers, admitted for a short stay (maximum 90 days in any 180-day period) in the Schengen Area. The EES should collect data (identity and travel document) and register entry and exit records (date and place of entry and exit) with a view to facilitating the border crossing of bona fide travelers at the same time as being able to identify overstayers. The EES will also record refusals of entry. The EES will replace the current system of manual stamping of passports." See "Entry/Exit System (EES)," *European Commission*, accessed May 27, 2020, https://ec.europa.eu/home-affairs/content/entryexit-system-ees_en.

59. "European Travel Information and Authorisation System (ETIAS)," briefing, *European Parliament*, October 18, 2018, accessed May 27, 2020, http://www.europarl.europa.eu/RegData/etudes/BRIE/2017/599298/EPRS_BRI(2017)599298_EN.pdf.

60. "Security Union: European Commission Welcomes the Final Adoption of the New European Criminal Records Information System on Convicted Third Country Nationals," press release, April 9, 2019, European Commission, http://europa.eu/rapid/press-release_IP-19-2018_en.htm.

61. "Security Union: European Commission Welcomes the Final Adoption."

62. "EU Votes to Create a Gigantic Biometrics Database," *Schengen Visa Info*, April 23, 2019, 2020, https://www.schengenvisainfo.com/news/eu-votes-to-create-a-gigantic-biometrics-database-despite-criticism/.

63. "EU Pushes to Link Tracking Databases," *Politico*, April 15, 2019 (updated April 16, 2019), accessed May 27, 2020, https://www.politico.eu/article/eu-pushes-to-link-tracking-databases/.

64. Thomas Nail explains in *Theory of the Border* (New York: Oxford University Press, 2016), 13 that his method is mainly materialist, as he "understands borders as regimes of concrete techniques and not primarily as ideas or knowledges that emerged independently from social and material conditions." The crucial word here, of course, is "independently."

65. Julien Jeandesboz, "An Analysis of the Commission Communications on Future Development of Frontex and the Creation of a European Border Surveillance System (Eurosur)," *Briefing Paper European Parliament* (Brussels: European Parliament, 2008), 6–7.

66. Gregory Feldman, *The Migration Apparatus* (Stanford, CA: Stanford University Press, 2012), 94.

67. According to Regulation (EU) No 1052/2013 of the European Parliament and of the Council of 22 October 2013, "The establishment of a European Border Surveillance System ('EUROSUR') is necessary in order to strengthen the exchange of information and the operational cooperation between national authorities of Member States as well as with the European Agency for the Management of Operational Cooperation at the External Borders of the Member States of the European Union established by Council Regulation (EC) No 2007/2004 (2) ('the Agency'). EUROSUR will provide those authorities and the Agency with the infrastructure and tools needed to improve their situational awareness and reaction capability at the external borders of the Member States of the Union ('external borders') for the purpose of detecting, preventing and combating illegal immigration and cross-border crime and contributing to ensuring the protection and saving of the lives of migrants."

68. "Opinion 4/2018 on the Proposal for Two Regulations Establishing a Framework for Interoperability between EU Large-Scale Information Systems," European Data Protection Supervisor, April 16, 2018, 30.

69. Stephan Dünnwald, "Europe's Global Approach to Migration Management: Doing Border in Mali and Mauritania," in *Externalizing Migration Management: Europe, North*

America and the Spread of 'Remote Control' Practices, ed. Ruben Zaiotti (London/New York: Routledge, 2016), 116–119.

70. Henriette Ruhrmann and David FitzGerald, "The Externalization of Europe's Borders in the Refugee Crisis, 2015–2016," San Diego, CA, University of California Center for Comparative Immigration Studies, 2017, 5, 22.

71. According to Luiza Bialasiewicz, in "Off-shoring and Out-sourcing the Borders of EUrope: Libya and EU Border Work in the Mediterranean," *Geopolitics* 17, no. 4 (2012): 844, this is where we can best "perceive that which Peter Sloterdijk (1994) has called the uniquely European process of 'translatio imperii.' It is at/through borders that the European space is constituted and selectively 'stretched'—or, to use Sloterdijk's terms, 'translated.'"

72. "Refugee Crisis Demands New Deal with Africa," News item, Government of the Netherlands, November 7, 2015, https://www.government.nl/latest/news/2015/11 /07/refugee-crisis-demands-new-deal-with-africa.

73. Narin Idriz, in "The EU-Turkey Deal in Front of the Court of Justice of the EU: An Unsolicited Amicus Brief," *T.M.C. Asser Institute for International & European Law*, Policy Brief 2017–03, accessed May 26, 2020, https://ssrn.com/abstract=3080838 discusses the relationship of the "Statement" with Article 218 of the Treaty on the Functioning of the European Union (TFEU).

74. Sergio Carrera, Leonhard den Hertog, and Marco Stefan, in "It Wasn't Me! The Luxembourg Court Orders on the EU-Turkey Refugee Deal," *CEPS Policy Insights* No. 2017/15, April 2017, 1, state that "the EU institutions purposefully—and unfortunately, successfully—circumvented the democratic and judicial checks and balances as aid down in the EU Treaties. . . . By choosing to conduct major policy decisions through press releases and refusing to take legal responsibility for the Statement, the EU institutions themselves jeopardize the Treaty-based framework that aims to ensure democratic rule of law and fundamental rights."

75. Ceccorulli and Lucarelli, "Securing the EU's Borders," 7.

76. Claudia Aradau and Jef Huysmans, "Critical Methods in International Relations: The Politics of Tools, Devices and Acts," *European Journal of International Relations* 20, no. 3 (2014): 604.

77. Thomas Spijkerboer, "Afterword: From the Iron Curtain to Lampedusa," in *Border Deaths: Causes, Dynamics and Consequences of Migration-Related Mortality*, ed. Paolo Cuttitta and Tamara Last (Amsterdam: Amsterdam University Press, 2020), 164.

78. FitzGerald, *Refuge beyond Reach*, 201–207.

79. See Paolo Cuttita, "Pushing Migrants Back to Libya, Persecuting Rescue NGOs: The End of the Humanitarian Turn (Part I)," accessed January 4, 2019, https://www .law.ox.ac.uk/research-subject-groups/centre-criminology/centreborder-criminologies

/blog/2018/04/pushing-migrants; and "Pushing Migrants Back to Libya, Persecuting Rescue NGOs: The End of the Humanitarian Turn (Part II)," accessed January 4, 2019, https://www.law.ox.ac.uk/research-subject-groups/centre-criminology/centreborder -criminologies/blog/2018/04/pushing-0.

80. Thomas Spijkerboer, "Afterword: From the Iron Curtain to Lampedusa," in *Border Deaths: Causes, Dynamics and Consequences of Migration-Related Mortality*, ed. Paolo Cuttitta and Tamara Last (Amsterdam: Amsterdam University Press, 2020), 164.

81. Institute for Race Relations (IRR), "Humanitarianism: The Unacceptable Face of Solidarity" (London: IRR, 2017).

82. Huub Dijstelbloem and William Walters, "Atmospheric Border Politics: The Morphology of Migration and Solidarity Practices in Europe," *Geopolitics*, 2019, DOI: 10.1080/14650045.2019.1577826.

83. "Migration Data Management, Intelligence and Risk Analysis," IOM, accessed May 28, 2020, https://www.iom.int/migration-data-management-intelligence-and -risk-analysis.

84. "IOM and Biometrics," IOM, November 2018, https://www.iom.int/sites/default /files/our_work/DMM/IBM/iom_and_biometrics_external_info_sheet_novem- ber_2018.pdf.

85. See Philippe Frowd, *Security at the Borders. Transnational Practices and Technologies in West Africa* (Cambridge: Cambridge University Press, 2018); and "The Promises and Pitfalls of Biometric Security Practices in Senegal," *International Political Sociology* 11, no. 4 (2017).

86. See Frank Schipper and Johan Schot, "Infrastructural Europeanism, or the Proj- ect of Building Europe on Infrastructures: An Introduction," *History and Technology* 27, no. 3 (2011).

87. For the historiographical meaning of the "laboratory" notion and how it was used by Schengen officials and EU officials as a metaphor, see William Walters and Jens Henrik Haahr. *Governing Europe: Discourse Governmentality and European Integra- tion* (New York: Routledge, 2005), 144; Zaiotti, *Cultures of Border Control*, 75–76; and Emek Uçarer, "Justice and Home Affairs," in *European Union Politics*, ed. Michelle Cini and Nieves Pérez-Solórzano Borragán (Oxford: Oxford University Press, 2019). Jörg Monar, in "The Dynamics of Justice and Home Affairs: Laboratories, Driving Factors and Costs," *Journal of Common Market Studies* 39, no. 4 (2001): 750, states that "the Schengen members never failed to emphasize that its role was to be that of a 'laboratory' for EC policymaking on the complete implementation of free move- ment and all related compensatory justice and home affairs measures."

88. See Bruno Latour, "Give Me a Laboratory and I Will Raise the World," in *Sci- ence Observed. Perspectives on the Social Study of Science*, ed. Karin Knorr-Cetina and Michael Mulkay (London: SAGE, 1983), 141–170.

89. Jan Hendrik Passoth and Nicholas Rowland, "Actor-Network State: Integrating Actor-Network Theory and State Theory," *International Sociology* 25, no. 6 (2010): 832.

90. Broeders and Dijstelbloem, "The Datafication of Mobility and Migration Management."

91. Quoted in Zaiotti, *Cultures of Border Control*, 76.

92. As stated in Article 77 of Chapter 2 of the 2012 Consolidated Version of the Treaty on the Functioning of the European Union:

> 1. The Union shall develop a policy with a view to:
>   (a) ensuring the absence of any controls on persons, whatever their national-
>   ity, when crossing internal borders;
>   (b) carrying out checks on persons and efficient monitoring of the crossing of
>   external borders;
> AND:
>   (c) the gradual introduction of an integrated management system for exter-
>   nal borders

Article 77 continues:

> 2. For the purposes of paragraph 1, the European Parliament and the Council,
>   acting in accordance with the ordinary legislative procedure, shall adopt mea-
>   sures concerning:
>   (a) the common policy on visas and other short-stay residence permits;
>   (b) the checks to which persons crossing external borders are subject;
>   (c) the conditions under which nationals of third countries shall have the
>   freedom to travel within the Union for a short period;
>   (d) any measure necessary for the gradual establishment of an integrated
>   management system for external borders;
>   (e) the absence of any controls on persons, whatever their nationality, when
>   crossing internal borders.
>
> 3. If action by the Union should prove necessary to facilitate the exercise of the
>   right referred to in Article 20(2)(a), and if the Treaties have not provided the
>   necessary powers, the Council, acting in accordance with a special legisla-
>   tive procedure, may adopt provisions concerning passports, identity cards,
>   residence permits or any other such document. The Council shall act unani-
>   mously after consulting the European Parliament.
>
> 4. This Article shall not affect the competence of the Member States concerning
>   the geographical demarcation of their borders, in accordance with interna-
>   tional law.

93. Stuart Elden, *The Birth of Territory* (Chicago: University of Chicago Press, 2013).

94. For instance, Nick Vaughan-Williams, in "Borderwork beyond Inside/Outside? Frontex, the Citizen-Detective and the War on Terror," *Space and Polity* 12, no. 1 (2008):

77, argues, "As a control on the movement of subjects into and within Europe, practices of surveillance can be read as a form of bordering. Increasingly, such a control takes place in spaces that cannot be readily identified as either internal or external border sites in a simplistic sense."

95. As Wendy Brown states in *Walled States, Waning Sovereignty* (Cambridge, MA: MIT Press, 2011), 82, "these terms 'inside' and 'outside' do not necessarily correspond to nation-state identity or fealty, that is, where otherness and difference are detached from jurisdiction and membership."

96. In *Hollow Land: Israel's Architecture of Occupation* (New York: Verso, 2012), 4–6, Eyal Weizman speaks of an elastic geography of myriad actors and dispersed spatial organization. He concludes that the architecture of borders "cannot be understood as the material embodiment of a unified political will or as the product of a single ideology." Instead, "the architecture of the frontier could not be said to be simply 'political' but rather 'politics in matter.'" Weizman explains the architecture of border politics can be thought of in two ways. It can be used to read specific forms of border politics "in the way social, economic, national, and strategic forces solidify into the organization, form, and ornamentation of homes, infrastructure, and settlements." But it can also be employed "as a conceptual way of understanding political issues as constructed realities."

97. The political and material geography of borders is linked to what Louise Amoore, in "Cloud Geographies. Computing, Data, Sovereignty," *Security Dialogue* 42, no. 1 (2018): 4–24 calls the cloud geography of borders.

98. Rob Walker in *Inside/Outside: International Relations as Political Theory* (Cambridge: Cambridge University Press, 1993), 6 expresses his dissatisfaction with theories that "simply take historically specific-modern-ontological options as a given." His conceptualization of the inside and outside of states in a shifting global world order has not left the study of borders unaffected. As John Allen argues *Topologies of Power. Beyond Territory and Network* (Oxford and New York: Routledge, 2016), 136–137 "the idea of a 'blurred' inside and outside . . . seems to miss the point somewhat." Instead, "a variety of modes of exclusion and inclusion may be exercised" with a number of shades of gray in between, as Huub Dijstelbloem and Dennis Broeders argue in "Border Surveillance, Mobility Management and the Shaping of Non-publics in Europe," *European Journal of Social Theory* 18, no. 1 (2014): 21–38.

99. As Reece Jones argues in *Border Walls: Security and the War on Terror in the United States, India and Israel* (Chicago: Zed Books, 2012), 171, "The construction of a barrier on the border simultaneously legitimizes and intensifies these other exclusionary practices of the sovereign state. It legitimizes exclusion by providing a material manifestation of the abstract idea of sovereignty, which brings the claim of territorial difference into being. The barrier also intensifies these exclusionary practices, because once the boundary is marked 'the container' of the state takes form, the

perception of difference between the two places becomes stronger. By performing sovereign control, the state simultaneously reifies authority over that territory and defines the limits of the people that belong there." Of particular interest here is Jones's remark here that "once the boundary is marked 'the container' of the state takes form."

100. Linnet Taylor in "No Place to Hide? The Ethics and Analytics of Tracking Mobility Using Mobile Phone Data," *Environment and Planning D: Society and Space* 34, no. 2 (2016): 323 says, "Big data enables us to map people and movement without necessarily mapping land. The people are the territory. Thus, if the aim of much GIS work has historically been to establish claims to land and to govern people, the new data technologies are in comparison more remote: they allow the viewer to track, often in real time, and to influence. Particularly with regard to the transgression of state boundaries involved in irregular migration, as will be explored here, the new data from digital traces lend themselves to a post-Westphalian politics of influence and indirect action."

Chapter 3

1. The intimate relation between politics and technology resonates at the level of concrete border technology devices. As Karolina Follis points out in "Vision and Transterritory: The Borders of Europe," *Science, Technology, and Human Values* 42, no. 6 (2017): 1007, such devices "are not just paraphernalia or elements of the border spectacle . . . Rather, they are designed effectively to project power beyond the physical boundaries of sovereign territory."

2. Bruno Latour, *We Have Never Been Modern* (Cambridge, MA: Harvard University Press, 1993), 67; Peter Sloterdijk, *Not Saved: Essay after Heidegger* (Cambridge, UK: Polity Press, 2017), 384.

3. The discussion of Latour and Sloterdijk in this chapter builds in part on chapter 4 "The technological atmosphere" ["De technologische atmosfeer"], 89–108 of Huub Dijstelbloem, *The House of Argus: The Watchful Eye in Democracy* [*Het Huis van Argus: De Wakende Blik in de Democratie*] (Amsterdam: Boom, 2016), but it has been thoroughly expanded, updated, and revised.

4. Willem Schinkel and Liesbeth Noordegraaf-Eelens, *In Medias Res: Peter Sloterdijk's Spherological Poetics of Being* (Amsterdam: Amsterdam University Press, 2011), 7.

5. "I take an interest in proving that human beings are no mono-elementary creatures. Whoever considers them in this way, is simply wrong. Almost all anthropology is suffering from a mono-elementary bias. It interprets us as creatures who in the end can only exist in one element, that is to say, on the mainland, in the so-called real. Against this tendency, I have been developing a theory of moves, a theory of transitions between elements and situations" Peter Sloterdijk and H.-J. Heinrichs,

Die Sonne und der Tod: Dialogische Untersuchungen (Frankfurt am Main: Suhrkamp, 2001), 336 quoted in René Ten Bos, "Towards an Amphibious Anthropology: Water and Peter Sloterdijk," *Environment and Planning D: Society and Space* 27 (2009): 79.

6. Peter Sloterdijk, *Foams* (Cambridge, MA: MIT Press, 2016), 107.

7. Bruno Latour, *Facing Gaia: Eight Lectures on the New Climate Regime* (Oxford, UK: Wiley-Blackwell, 2017), 88.

8. Henk Oosterling, "Dasein als Design," *De Groene Amsterdammer* 130, no. 14 (2009): 32.

9. Peter Sloterdijk, *Foams* (Cambridge, MA: MIT Press, 2016), 23.

10. Sloterdijk, *Foams*, 25.

11. John Gray's "Blowing Bubbles," *New York Review of Books*, October 12, 2017, critically reviews Sloterdijk's work through this lens.

12. See Reviel Netz, *Barbed Wire: An Ecology of Modernity* (Middletown, CT: Wesleyan University Press, 2004).

13. "How Can Concertina Wire Supplier Help You?" *Hebei Wanxiang Concertina Wire Company*, accessed May 28, 2020, http://www.concertina-wire.org/.

14. In *Borderlands/La Frontera: The New Mestiza* (San Francisco: Aunt Lute Books, 2012), 24, Gloria Anzaldúa refers to the barbwired US-Mexico border as a "steel curtain" that creates an "open wound."

15. "Hungary Builds New High-Tech Border Fence—with Few Migrants in Sight," *Reuters*, March 2, 2017, https://www.reuters.com/article/us-europe-migrants-hungary -fence/hungary-builds-new-high-tech-border-fence-with-few-migrants-in-sight-idUS KBN1692MH.

16. "'A Bloody Method of Control': The Struggle to Take Down Europe's Razor Wire Walls," *The Guardian*, May 13, 2020, https://www.theguardian.com/global-development /2020/may/13/a-bloody-method-of-control-the-struggle-to-take-down-europes-razor-wire -walls.

17. What began as chance encounters early in the millennium were followed by Latour writing a foreword to the French translation of Sloterdijk's *Rules for the Human Zoo: A Response to Heidegger's Letter on Humanism* [*Regeln für den Menschenpark. Ein Antwortschreiben zu Heideggers Brief über den Humanismus*] (Frankfurt: Suhrkamp, 2008). In turn, Sloterdijk provided a chapter for the catalog of the exhibition *Making Things Public*. Thereafter, the two made frequent references to each other's work and appeared together, such as at Harvard and Columbia University in 2009. They even spoke in each other's honor, like when Latour was awarded the Siegfried Unseld Prize in 2008. In 2017, Latour was one of the speakers at a symposium to celebrate Sloterdijk's seventieth birthday. (See Thomas Meany, "A Celebrity Philosopher

Explains the Populist Insurgency," *The New Yorker*, February 26, 2018, https://
www.newyorker.com/magazine/2018/02/26/a-celebrity-philosopher-explains-the
-populist-insurgency.) Over the years, each has expressly sought a rapprochement
with the other's work. For Sloterdijk, this was expressed in the *Spheres* trilogy. For
Latour, it can be seen in *Making Things Public: Atmospheres of Democracy*, edited by
Latour and Peter Weibel (Cambridge, MA: MIT Press, 2005), which he put together
for the exhibition of the same name; and in *Facing Gaia*.

18. See Latour, *We Have Never Been Modern*, and Latour, *Reassembling the Social: An
Introduction to Actor-Network-Theory*.

19. See Latour, *We Have Never Been Modern*.

20. For the study of borders and security, William Walters, in "Drone Strikes,
Dingpolitik and Beyond: Furthering the Debate on Materiality and Security," *Security Dialogue* 45, no. 2 (2014): 101–118, has advanced the notion of *Dingpolitik* to
analyze Human Rights Watch's investigation of Gaza civilians allegedly killed by
Israeli drone-launched missiles in 2008–2009. For Walters, "*dingpolitik* offers some
crucial guidelines for an understanding of how material things become entangled in
disputes, and how political controversy is mediated, shaped and channelled by the
affordance of things. . . . *Dingpolitik* identifies multiple assemblies, not just the ones
that convene in parliamentary buildings." Walters concludes that Latour's ontopoliti-
cal repertoire is useful not only to point to the importance of materialities of all sorts,
but also to conceive the coming-into-being of issues and publics via technologies
and how situations become visible or invisible. *Dingpolitik* proves to be an intriguing
notion to "catch" publics via technologies. See also William Walters and Anne-Marie
D'Aoust, "Bringing Publics into Critical Security Studies: Notes for a Research Strategy," *Millennium: Journal of International Studies* 44, no. 1 (2015): 45–68.

21. Sloterdijk, *Foams*, 193.

22. Sloterdijk, *Foams*, 193.

23. Sloterdijk, *Foams*, 201.

24. Bruno Latour, "Gabriel Tarde and the End of the Social," in *The Social in
Question. New Bearings in History and the Social Sciences*, ed. Patrick Joyce, 117–132
(London: Routledge, 2001).

25. Sloterdijk, *Foams*, 567.

26. Sloterdijk, *Foams*, 602.

27. Sloterdijk, *Foams*, 565.

28. Sjoerd van Tuinen, *Sloterdijk. Binnenstebuiten denken* [*Thinking Inside Out*]
(Kampen, Netherlands: Klement/Pelckmans, 2004).

29. Latour, *The Pasteurization of France*, 229.

30. See Sheila Jasanoff, "Future Imperfects," in *Dreamscapes of Modernity*, ed. Sheila Jasanoff and Sang Hyun Kim (Chicago: University of Chicago Press, 2015), 1–34.

31. Jasanoff, in "Future Imperfects," *Dreamscapes of Modernity*, hastens to say that this has little to do with his expertise because he is too knowledgeable for that.

32. Jasanoff, "Future Imperfects," 17–18.

33. As Latour explains in "Why Has Critique Run out of Steam? From Matters of Fact to Matters of Concern," *Critical Inquiry* 30, no. 2 (2004): 245–246, "What is presented here is an entirely different attitude than the critical one, not a flight into the conditions of possibility of a given matter of fact, not the addition of something more human that the inhumane matters of fact would have missed, but, rather, a multifarious inquiry launched with the tools of anthropology, philosophy, metaphysics, history, sociology to detect how many participants are gathered in a thing to make it exist and to maintain its existence."

34. Sloterdijk's and Latour's works are examples of immanent philosophical thinking. Latour builds on the philosophies of Baruch Spinoza and Gilles Deleuze. Deleuze himself called Spinoza the "Prince of Philosophers," in view of his unparalleled systematicity in his development of the notion of immanence. Latour and Sloterdijk are also indebted to Nietzsche. It was Nietzsche who, with his declaration that "God is dead," drew a definitive line through transcendental thinking. With the death of God, he ended thinking from outside, the idea of an external point of view, and an external reference point that is absent but still determines our existence and our thinking. This death certificate of the transcendental affects both the conditions that are put to thinking and the claims that can be made on the basis of that thinking. This life termination of the transcendental resonates in the impossibility of the "god trick of seeing everything from nowhere," as Donna Haraway points out in "Situated Knowledges: The Science Question in Feminism and the Privilege of Partial Perspective," *Feminist Studies* 14, no. 3 (1988): 581.

35. Latour, *Facing Gaia*, 88.

36. Latour, *Facing Gaia*, 91.

37. Latour, *Facing Gaia*, 116.

38. In particular, see Eric Voegelin's *The New Science of Politics* (Chicago: University of Chicago Press, 1987).

Chapter 4

1. Peter Sloterdijk, *Foams* (Cambridge, MA: MIT Press, 2016), 165.

2. "Dat is wat ik bewaak, de geest van Schiphol," interview with Jan Benthem in *Het Parool*, July 10, 2016, https://www.parool.nl/nieuws/dat-is-wat-ik-bewaak-de-geest-van -schiphol~b8400ea1/.

3. "The Interior of Schiphol: Teamwork on a Never-Ending Work in Progress," *Design History*, accessed May 28, 2020, http://www.designhistory.nl/2016/the-interior-of -schiphol-teamwork-on-a-never-ending-work-in-progress/.

4. Henk van Houtum, "Human Blacklisting: The Global Apartheid of the EU's External Border Regime," *Environment and Planning D: Society and Space* 28, no. 6 (2010): 957; and Thomas Spijkerboer, "The Global Mobility Infrastructure: Reconceptualising the Externalisation of Migration Control," *European Journal of Migration and Law* 20, no. 4 (2018): 452–469.

5. David Scott FitzGerald, *Refuge beyond Reach. How Rich Democracies Repel Asylum Seekers* (Oxford: Oxford University Press, 2019), 221.

6. See, for instance, Avishai Margalit, *On Compromise and Rotten Compromises* (Princeton, NJ: Princeton University Press, 2009), which distinguishes between acceptable and rotten compromises.

7. See Amy Gutmann and Dennis Thompson, *The Spirit of Compromise. Why Governing Demands It and Campaigning Undermines It* (Princeton, NJ: Princeton University Press, 2010), which describes the polarization of partisan politics in the United States and the decline of the noble art of compromise.

8. This approach is informed by an argument by Luc Boltanski and Laurent Thévenot in *On Justification: Economies of Worth* (Princeton, NJ: Princeton University Press, 2006). Following Boltanski and Thévenot, neither a general theory of justice to evaluate the value of compromises nor a theory of deliberative democracy to study their making and unmaking must be applied.

9. Boltanski and Thévenot, *On Justification*, 278.

10. As such, "airports make up central nodes in the critical infrastructure of globalization, where the circulation of high quantities of goods, persons and capital are managed," according to Peer Schouten in "Security as Controversy: Reassembling Security at Amsterdam Airport," *Security Dialogue* 45, no. 1. (2014): 24.

11. The analysis of Schiphol International Airport in this chapter partly builds on chapter 5 "Designed Space" ["Ontworpen ruimte"] of Huub Dijstelbloem, *The House of Argus: The Watchful Eye in Democracy* [*Het Huis van Argus: De Wakende Blik in de Democratie*] (Amsterdam: Boom, 2016), 83–94, but has been thoroughly expanded, updated, and revised.

12. The concept of AirportCity "captures the experimental development of the airport into a unique semi-public space open 24/7 that does not substitute a city center but creates a new image of 'cityness' born out of a juxtaposition of functions and a very peculiar diversity of users," Anna Nikolaeva, "Designing Public Space for Mobility: Contestation, Negotiation and Experiment at Amsterdam Airport Schiphol," *Tijdschrift voor Economische en Sociale Geografie* 103, no. 5 (2012): 549.

13. Bart de Jong, *The Airport Assembled. Rethinking Planning and Policy Making of Amsterdam Airport Schiphol by Using the Actor-Network theory* (PhD thesis, Utrecht University, 2012), 27.

14. Mark Salter (Ed.), *Politics at the Airport* (Minneapolis: University of Minnesota Press, 2008).

15. Stephen Graham, "Flowcity: Networked Mobilities and the Contemporary Metropolis," *Journal of Urban technology* 9, no. 1 (2002): 4–11.

16. Kenneth Frampton, *A Genealogy of Modern Architecture. Comparative Critical Analysis of Built Form* (Zurich: Lars Müller Publishers, 2015).

17. John Kasarda and Greg Lindsay, *Aerotropolis. The Way We'll Live Next* (New York: Farrar, Straus and Giroux, 2011).

18. See Francisca Grommé, *Governance by Pilot Projects: Experimenting with Surveillance in Dutch Crime Control* (PhD thesis, University of Amsterdam, 2015) and Marijn Hoijtink, *Securing the European "Homeland": Profit, Risk, Authority* (PhD thesis, University of Amsterdam, 2016).

19. Étienne Balibar argues, in *We, the People of Europe? Reflections on Transnational Citizenship* (Princeton, NJ: Princeton University Press, 2004), that "the system of identity verifications . . . [allows] a triage of travelers admitted to and rejected from a given national territory. For the mass of humans today, these are at the most decisive borders, but they are no longer 'lines': instead they are *detention zones* and *filtering systems* such as those located in the center or on the periphery of major international airports" (italics in the original).

20. John Allen, in *Topologies of Power. Beyond Territory and Network* (Oxford and New York: Routledge, 2016), 130, argues that it is "the growth and sophistication of border security technologies, biometrics and pre-screening through data-mining in particular, designed to sort out the 'safe' from the 'risky' population, that has helped to anchor the idea that border controls and checks are now pervasive throughout society."

21. Mark Dierikx, Johan Schot, and Ad Vlot, "Van uithoek tot knooppunt: Schiphol," in *Techniek in Nederland in de Twintigste Eeuw, Deel V: Transport en Communicatie*, ed. Johan Schot and Harry Lintsen (Zutphen, Netherlands: Walburg Pers, 2002).

22. Gillian Fuller, "Welcome to Windows 2.1: Motion Aesthetics at the Airport," in *Politics at the Airport*, ed. Mark B. Salter (Minneapolis: University of Minnesota Press, 2008).

23. Francis D. K. Ching, *Form, Space, and Order* (New York: John Wiley & Sons, 1996); and David Harvey, "The Political Economy of Public Space," in *The Politics of Public Space*, ed. Setha Low and Neil Smith (New York: Routledge, 2005), 17–34.

24. Nikolaeva, "Designing Public Space for Mobility," 542–554.

25. Ewald Engelen, Julie Froud, Sukhdev Johal, Angelo Salento, and Karel Williams, "How Cities Work: A Policy Agenda for the Grounded City," *Cresc Working Paper Series*, Working Paper 141, 2016, 2.

26. Peter Sloterdijk, *Globes* (Cambridge, MA: The MIT Press, 2014).

27. Sze Tsung Leong, "Gruen Urbanism," in *Harvard Design School Guide to Shopping*, ed. Chuihua Judy Chung, Jeffrey Inaba, Rem Koolhaas and Sze Tsung Leong (Cambridge, MA/Cologne, Germany: Harvard Design School/Taschen, 2001); and Sze Tsung Leong and Srdjan Jovanovic Weiss, "Air Conditioning," *Harvard Design School Guide to Shopping*.

28. Frederike Huygen, "The Interior of Schiphol: Teamwork on a Never-Ending Work in Progress," *Design History*, accessed May 28, 2020, http://www.designhistory .nl/2016/the-interior-of-schiphol-teamwork-on-a-never-ending-work-in-progress. This text was originally published in Dutch in *Flow: Het Schiphol van Nel Verschuuren 1968–2005* (Schiphol Group, 2006).

29. Frederike Huygen, "The Interior of Schiphol: Teamwork on a Never-Ending Work in Progress," *Design History*, accessed May 28, 2020, http://www.designhistory .nl/2016/the-interior-of-schiphol-teamwork-on-a-never-ending-work-in-progress. This text was originally published in Dutch in *Flow: Het Schiphol van Nel Verschuuren 1968–2005* (Amsterdam: Schiphol Group, 2006).

30. See Louise Amoore, "Lines of Sight: On the Visualization of Unknown Futures," *Citizenship Studies* 13, no. 1 (2009): 17–30; and Jonathan Crary, *Suspensions of Perception: Attention, Spectacle, and Modern Culture* (Cambridge, MA: MIT Press, 1999). Chapter 7 of this book will offer a broader discussion of this notion.

31. Louise Amoore, *The Politics of Possibility: Risk and Uncertainty beyond Probability* (Durham, NC: Duke University Press, 2013), 100.

32. Peter Sloterdijk, *Globes* (Cambridge, MA: MIT Press, 2014).

33. Rem Koolhaas and Bruce Mau. *S, M, L, XL* (Rotterdam, Netherlands, and New York: 010 Publishers/ Monacelli Press, 1995); Maarten Hajer, "The Generic City," *Theory, Culture & Society* 16, no. 4 (1999): 137–144.

34. J. G. Ballard, "Airports," *The Observer*, September 14, 1997. Quoted in Will Self, "The Frowniest Spot on Earth," *London Review of Books* 33, no. 9 (2011): 10.

35. Self, "The Frowniest Spot on Earth," *London Review of Books* 33, no. 9 (2011): 10–11.

36. Mark Augé, *Non-Places* (London/New York: Verso, 2008).

37. Nikolaeva, "Designing Public Space for Mobility," 542.

38. Augé, *Non-Places*.

39. Rachel Hall, *The Transparent Traveler. The Performance and Culture of Airport Security* (Durham, NC/London: Duke University Press, 2015), 15.

40. Peer Schouten, "Security as Controversy: Reassembling Security at Amsterdam Airport," *Security Dialogue* 45, no. 1 (2014): 9.

41. Schouten, "Security as Controversy: Reassembling Security at Amsterdam Airport," 8.

42. See Louise Amoore, "Biometric Borders: Governing Mobilities in the War on Terror," *Political Geography* 25 (2006): 336–351; Amoore, *The Politics of Possibility*; and Marieke De Goede, Stephanie Simon, and Marijn Hoijtink, "Performing Preemption," *Security Dialogue* 45, no. 5 (2014): 411–422.

43. Amoore, *The Politics of Possibility*, 84, 102–103.

44. "Slimme camera op Schiphol spot afwijkend gedrag," *NRC Handelsblad*, September 11, 2014, https://www.nrc.nl/nieuws/2014/09/11/terreurbestrijding-slimme -camera-op-schiphol-spot-1420549-a1350699.

45. "Kick off Meeting," *Tresspass*, accessed June 2, 2020, https://www.tresspass.eu /news/kick-meeting.

46. "Schiphol als levend lab," *NRC Handelsblad*, November 29, 2019, https://www .nrc.nl/nieuws/2019/11/29/schiphol-als-levend-lab-a3982129.

47. As Gregory Feldman describes in *The Migration Apparatus* (Stanford, CA: Stanford University Press, 2012): 84, "the Privium Program not only creates frictionless travel for passengers with economic capital, but it also pads that experience with extra comfort."

48. James Bridle, "What They Don't Want You to See: The Hidden World of UK Deportation," *The Guardian*, January 27, 2015, https://www.theguardian.com/artanddesign /2015/jan/27/hidden-world-of-uk-deportation-asylum-seamless-transitions.

49. "Vreemdeling behandeld als crimineel," Jojanneke Spoor, accessed February 14, 2017, http://www.napnieuws.nl/2010/10/22/6074/.

50. According to Arjen Leerkes and Dennis Broeders, in "A Case of Mixed Motives? Formal and Informal Functions of Administrative Immigration Detention," *British Journal of Criminology* 50 (2010): 830–850, "it is noteworthy that the Dutch Expulsion Centers in Rotterdam and at Schiphol Airport were introduced under the banner of a government program that was called 'Towards a Safer Society.' . . . It seems that Dutch authorities increasingly use immigration detention (and criminal detention) for incapacitation purposes and not only as a measure of immigration policy."

51. "Apparently, somebody may be regarded as an 'illegal' even before a legal system had declared that they were attempting to remain on the soil of a European country without authorization by the state," according to Yolande Jansen, "Deportability and Racial Europeanization: The Impact of Holocaust Memory and Postcoloniality on the Unfreedom of Movement in and to Europe," in *The Irregularization of Migration in*

Contemporary Europe. Detention, Deportation, Drowning, ed. Yolande Jansen, Robin Celikates, and Joost de Bloois (London and New York: Rowman and Littlefield, 2015), 15–16.

52. Dutch Safety Board. *Fire at the Detention Centre Schiphol Oost* (The Hague: Dutch Safety Board, 2006), 170–174.

53. Giorgio Agamben, *Homo Sacer: Sovereign Power and Bare Life* (Stanford, CA: Stanford University Press, 1998).

54. See Lieven de Cauter, *De capsulaire beschaving. Over de stad in het tijdperk van de angst* (Rotterdam, Netherlands: NAI Uitgevers, 2009).

55. See "About JFKIAT," accessed November 10, 2020, https://www.jfkt4.nyc/about/about-jfkiat/.

56. In the opening lines of *Paris, Invisible City*, 2006, accessed May 28, 2020, http://www.bruno-latour.fr/node/343, Bruno Latour states that "we often tend to contrast real and virtual, hard urban reality and electronic utopias." As a corrective, he aims to show that real cities consist of all kinds of networks that gather and circulate information so as to connect the streets to the maps to the control centers. As a result, "no single control panel or synoptic board brings all these flows together in a single place at any one time. . . . No bird's eye view could, at a single glance, capture the multiplicity of these places. . . . There are no more panopticons than panoramas; only richly colored dioramas with multiple connections, criss-crossing wires under roads and pavements, along tunnels in the metro, on the roofs of sewers. . . . The total view is also, literally, the view from nowhere."

57. Luc Boltanski and Laurent Thévenot, *On Justification: Economies of Worth* (Princeton, NJ: Princeton University Press, 2006).

58. Jacques Rancière, *The Politics of Aesthetics: The Distribution of the Sensible* (London and New York: Continuum, 2004).

59. John Allen, *Topologies of Power: Beyond Territory and Network* (Oxford and New York: Routledge, 2016).

60. According to Bruno Latour, in *Facing Gaia: Eight Lectures on the New Climate Regime* (Oxford, UK: Wiley-Blackwell, 2017), 93, this picture is misleading, as "the notion of a globe and any global thinking entails the immense danger of unifying too fast what should be composed instead." For this reason, Mark Salter, in "The Global Visa Regime and the Political Technologies of the International Self: Borders, Bodies, Biopolitics," *Alternatives: Global, Local, Political* 31, no. 2 (2006): 167–189, describes the development of a global visa regime in terms of the construction of specific "political technologies." William Walters, in "Rezoning the Global: Technological Zones, Technological Work, and the (Un-)Making of Biometric Borders," in *The Contested Politics of Mobility: Borderzones and Irregularity*, ed. Vicky Squire

(London: Routledge, 2011), 51–73 likewise analyzes the making of biometric borders in terms of the development of "technological zones."

Chapter 5

1. "Address by President Donald Tusk to the European Parliament on the Latest European Council of 15 October 2015," European Council, October 27, 2015, http://www.consilium.europa.eu/en/press/press-releases/2015/10/27/pec-speech-ep/.

2. The International Organization on Migration (IOM) is now affiliated with the United Nations (UN), but it was born in 1951 out of the displacement of Western Europe following World War II, with the name of Provisional Intergovernmental Committee for the Movement of Migrants from Europe (PICMME).

3. European Commission, "The Hotspot Approach to Managing Exceptional Migratory Flows," Fact Sheet, September 8, 2015, https://ec.europa.eu/home-affairs/sites/homeaffairs/files/what-we-do/policies/european-agenda-migration/background-information/docs/2_hotspots_en.pdf.

4. Michela Ceccorulli and Sonia Lucarelli, "Securing the EU's Borders in the Twenty-First Century," in *EU Security Strategies: Extending the EU System of Security Governance*, ed. Spyros Economides and James Sperling (Abingdon, UK: Routledge, 2018), 7.

5. This notion of a compromise shares some similarities with a Leviathan that emerges out of the interactions between actors and technologies without being part of the negotiating parties. In this comparison, the relationship between the Leviathan and the subjects ought to be understood as in the analysis of Michel Callon and Bruno Latour, in "Unscrewing the Big Leviathan, or How Do Actors Macrostructure Reality and How Sociologists Help Them to Do So," *Advances in Social Theory and Methodology: Toward and Integration of Micro and Macro Sociologies*, ed. K. Knorr Cetina and A. Cicourel (London: Routledge and Kegan Paul, 1981), 299. In an analysis of the controversy between Electricity of France and Renault over the future of transportation and the choice between the gasoline-fueled engine and the electric car in the 1970s, Callon and Latour investigate the relationship between large-scale concepts, such as the industrial society, sustainability, and the energy transition, and micro-innovations such as the fuel cell. Instead of explaining the controversy in terms of a clash of interests, ideas, or conflicting technical solutions and reducing the debate to either a macro or micro level of analysis, Callon and Latour emphasize the relation between the two. They reject the view that the Leviathan, as a macro actor, is opposed to the micro actors, the subjects. Instead they consider the co-construction and the mutual establishment of these various actors as taking place via the emergence of networks. Ideas and concepts such as "sustainability" or "the industrial society" are not of a different category than micro-actors such as fuel cells and injection engines. Instead, they argue that "a macro actor . . . is a micro actor seated on black boxes."

6. As in chapter 6, the interviews were organized, conducted, transcribed, and translated by Ermioni Frezouli. For a specific analysis of the entanglement between surveillance and rescue at sea based on this research, see Huub Dijstelbloem, Rogier van Reekum, and Willem Schinkel, "Surveillance at Sea: The Transactional Politics of Border Control in the Aegean," *Security Dialogue* 48, no. 3 (2017): 224–240.

7. As Özgün Topak observes in "The Biopolitical Border in Practice: Surveillance and Death at the Greece–Turkey Borderzones," *Environment and Planning D: Society and Space* 32, no. 5 (2014): 815–833, "surveillance at the Aegean Sea has gradually intensified over the years. The classic strategy of patrolling the sea with various types of vessels and air units is increasingly combined with 'smarter' systems, such as radars, satellites, and coordination centers."

8. "EUROSUR: Protecting the Schengen External Borders—Protecting Migrants' Lives," European Commission, Brussels, November 29, 2013, http://europa.eu/rapid /press-release_MEMO-13-1070_en.htm.

9. As Julien Jeandesboz explains in "European Border Policing. EUROSUR, Knowledge, Calculation," *Global Crime* 18, no. 3 (2017): 256–285, EUROSUR includes "the monitoring, detection, identification, tracking, prevention and interception of unauthorized border crossings for the purpose of detecting, preventing and combating illegal immigration and cross-border crime and contributing to ensuring the protection and saving of the lives of migrants." The description is partly based on OJEU, Regulation (EU) No. 1052/2013, Article 2(1).

10. See Karin Knorr Cetina and Alex Preda, "The Temporalization of Financial Markets: From Network to Flow," *Theory, Culture & Society* 24, no. 7 (2007): 116. In addition, see Karin Knorr Cetina, "Scopic Media and Global Coordination: The Mediatization of Face-to-Face Encounters," in *Mediatization of Communication*, ed. Knut Lundby (Berlin: de Gruyter, 2014).

11. The term "infrastructural Europeanism" was coined by Frank Schipper and Johan Schot in "Infrastructural Europeanism, or the Project of Building Europe on Infrastructures: An Introduction," *History and Technology* 27, no. 3 (2011).

12. Paul Edwards, "Meteorology as Infrastructural Globalism," *Osiris* 21: 229–250.

13. Frank Schipper and Johan Schot, "Infrastructural Europeanism, or the Project of Building Europe on Infrastructures: An Introduction," *History and Technology* 27, no. 3 (2011): 246.

14. As Dennis Duez and Rocco Bellanova argue in "The Making (Sense) of EUROSUR: How to Control the Sea Borders?" in *EU Borders and Shifting Internal Security: Technology, Externalization and Accountability*, ed. Raphael Bossong and Helena Carrapico (New York: Springer, 2016), 24, "The making of the sea borders operated by EUROSUR is, first and foremost, an effort to make sense of a disparate and heterogeneous ensemble of elements. This controlled space does not only concentrate on and

encompass potential migrants, small vessels of smugglers, and international networks of criminals. This kind of border surveillance is also, at the same time, and somehow prominently, an effort to understand and maximize the potential use of different elements—radars, national authorities, boats, information analysis systems, etc.—already deployed for border surveillance. Hence, the setup of a surveillance system is both a matter of material and symbolic controls, and a continuous effort of mise-en-discours of protean elements. It is an attempted and continuous mustering of things, people, information, institutions, programs, and research."

15. See Huub Dijstelbloem, "Migration Tracking Is a Mess," *Nature* 543, no. 2 (2017): 32–34.

16. Regulation (EU) No 1052/2013 of the European Parliament and of the Council, October 22, 2013.

17. Lucy Suchman, "Situational Awareness: Deadly Bioconvergence at the Boundaries of Bodies and Machines," *Media Tropes* 5, no. 1 (2015): 1.

18. Geoffrey A. Boyce, "The Rugged Border: Surveillance, Policing and the Dynamic Materiality of the US/Mexico Frontier," *Environment and Planning D: Society and Space* 34, no. 2 (2016): 246.

19. Martina Tazzioli, "Eurosur, Humanitarian Visibility, and (Nearly) Real-Time Mapping in the Mediterranean," *ACME: An International Journal for Critical Geographies* 15, no. 3 (2016): 562, 563, 566.

20. European Parliament Resolution of 18 December 2008 on the Evaluation and Future Development of the Frontex Agency and of the European Border Surveillance System (Eurosur) (2008/2157(INI)) (2010/C 45 E/08).

21. See Ruben Andersson, *Illegality, Inc.: Clandestine Migration and the Business of Bordering Europe* (University of California Press, 2014), 87–88.

22. See Julien Jeandesboz, in "European Border Policing. EUROSUR, Knowledge, Calculation," *Global Crime* 18, no. 3 (2017): 257, 275.

23. Concrete practices in the area of border control are often accompanied by visions of the future. The report that advised setting up SIVE—the so-called Civipol study—already suggested generalizing the use of "SIVE-type" systems. Indicators of the future orientation of the study include the use of terms like "guiding images." The priority areas for deploying such systems included the "southern contact zone between Spain and Morocco" (essentially the Canary Islands region), the "contact zone between Italy, Tunisia and Libya," and the "contact zone between Greece and Turkey." The report further stressed the possibility of using what it calls "new technologies," including the Galileo satellite network and unmanned aerial vehicles. See Jeandesboz, "European Border Policing. EUROSUR, Knowledge, Calculation," 274.

24. See James Gibson, *The Ecological Approach to Visual Perception* (Boston: Houghton Mifflin, 1979), and Tim Ingold, *Being Alive: Essays on Movement, Knowledge and Description* (London: Routledge, 2011).

25. Rogier van Reekum and Willem Schinkel, "Drawing Lines, Enacting Migration: Visual Prostheses of Bordering Europe," *Public Culture* 29, nos. 1/81 (2017): 34.

26. William Walters, "Live Governance, Borders, and the Time–Space of the Situation: EUROSUR and the Genealogy of Bordering in Europe," *Comparative European Politics* 15, no. 5 (2017): 794–817, begins with a quote from Peter Sloterdijk. The quote says that the term "live" allows us "to participate in events elsewhere." See Peter Sloterdijk, *Selected Exaggerations: Conversations and Interviews 1993–2012*, ed. B. Klein (Cambridge, UK: Polity, 2016).

27. Dijstelbloem et al., "Surveillance at Sea."

28. The classic reference is Bruno Latour's "Give Me a Laboratory and I Will Raise the World," in *Science Observed. Perspectives on the Social Study of Science*, ed. Karin Knorr-Cetina and Michael Mulkay (London: SAGE, 1983). Elaborating on that view, Jan Hendrik Passoth and Nicholas Rowland, in "Actor-Network State: Integrating Actor-Network Theory and State Theory," *International Sociology* 25, no. 6 (2010): 818–841, have conceptualized states as laboratories. Likewise, Mike Bourne, Heather Johnson, and Debbie Lisle, in their study on security devices "Laboratizing the Border: The Production, Translation and Anticipation of Security Technologies," *Security Dialogue* 46, no. 4 (2015): 307–325, compare and connect laboratory sites and border sites. Martina Tazzioli's "The Circuits of Financial-Humanitarianism in the Greek Migration Laboratory," September 25, 2017, https://www.law.ox.ac.uk/research-subject-groups /centre-criminology/centreborder-criminologies/blog/2017/09/circuits, describes the first Refugee Cash Assistance program funded by the European Union at the island of Lesbos as a "laboratory of experimentation." Glenda Garelli and Martina Tazzioli in "Choucha beyond the Camp: Challenging the Border of Migration Studies," in *The Borders of "Europe": Autonomy of Migration, Tactics of Bordering*, ed. Nicholas De Genova (Durham, NC/London: Duke University Press, 2017), 167, envision a specific refugee camp, Choucha in Tunisia, as a laboratory for a humanitarian border. Georgios Glouftsios, in "Designing Digital Borders: The Visa Information System (VIS)," *Technology and Agency in International Relations*, ed. Marijn Hoijtink and Matthias Seele (London/New York: Routledge, 2019), 167, describes the Visa Information System as a "laboratory."

29. See Ian Hacking's *Representing and Intervening: Introductory Topics in the Philosophy of Natural Science* (Cambridge, Cambridge University Press, 1983).

30. Bruno Latour, *Science in Action* (Cambridge, MA.: Harvard University Press, 1987), 215–257.

31. Interview with the director of Greece's Sea Border Protection Department, Piraeus, September 16, 2014.

32. John Torpey, *The Invention of the Passport. Surveillance, Citizenship and the State* (Cambridge: Cambridge University Press, 1998), 4–13.

33. Louise Amoore, *The Politics of Possibility: Risk and Uncertainty Beyond Probability* (Durham, NC: Duke University Press, 2013), 9, 81.

34. Amoore, in *The Politics of Possibility*, 102–103, argues, "As a mosaic of life signatures, the contemporary border is not merely a site of technology, where bodies become inscribed with code, but rather it becomes the sovereign enactment of possibility par excellence."

35. Interview with the union representative of the Hellenic Coast Guard at Chios, September 9, 2014.

36. Interview with the Lesbos coast guard commander, Lesvos, February 25, 2016.

37. Dijstelbloem et al., "Surveillance at Sea," 231.

38. Dijstelbloem et al., "Surveillance at Sea," 232.

39. Interview with the union representative of the Hellenic Coast Guard at Chios, September 9, 2014.

40. Martina Tazzioli, "Spy, Track and Archive: The Temporality of Visibility in Eurosur and Jora," *Security Dialogue* 49, no. 4 (2018): 274.

41. Martina Tazzioli, "Spy, Track and Archive: The Temporality of Visibility in Eurosur and Jora," *Security Dialogue* 49, no. 4 (2018): 276, 281.

42. "Jean Asselborn and Dimitris Avramopoulos, during a Visit to Greece, Take Stock of the Implementation of the First 'Hotspot' on the Island of Lesbos," *Presidency of the Council of the European Union*, October 10, 2015, http://www.eu2015lu .eu/en/actualites/articles-actualite/2015/10/16-asselborn-avramopoulos-lesbos /index.html.

43. See, for instance, *IOM Greece*, accessed May 28, 2020, https://greece.iom.int/en/iom -greece.

44. Charalambos Kasimis, "Illegal Immigration in the Midst of Crisis," *Migration Information Source*, March 8, 2012, 8–10, http://www.migrationpolicy.org/article/greece -illegal-immigration-midst-crisis.

45. European Commission, "The Hotspot Approach to Managing Exceptional Migratory Flows."

46. Tazzioli, "Spy, Track and Archive," 280.

47. Silvan Pollozek and Jan Hendrik Passoth, "Infrastructuring European Migration and Border Control: The Logistics of Registration and Identification at Moria Hotspot," *EPD: Society and Space* 37, no. 4 (2019): 3, 15.

48. Annalisa Pelizza, "Processing Alterity, Enacting Europe: Migrant Registration and Identification as Co-construction of Individuals and Polities," *Science, Technology, & Human Values* 25, no. 2 (2019): 9.

49. Antonis Vradis, Evie Papada, Joe Painter, and Anna Papoutsi state in *New Borders: Hotspots and the European Migration Regime* (London: Pluto Press, 2019), 8, "Once an area is declared a hotspot, the European Asylum Support Office (EASO), Frontex, Europol and Eurojust come in to assist member states to swiftly identify, register and fingerprint incoming migrants. What is envisioned by the EC is that the four agencies will support member-state authorities in the registration, identification and removal of apprehended migrants (using Frontex), the registration of asylum claims, the preparation of succesful relocation claimants (by EASO) and the investigation and subsequent prosecution of crimes (by Europol and Eurojust)."

50. "Jean Asselborn and Dimitris Avramopoulos, during a Visit to Greece, Take Stock of the Implementation of the First 'Hotspot' on the Island of Lesbos," *Presidency of the Council of the European Union*, October 10, 2015, http://www.eu2015lu.eu/en /actualites/articles-actualite/2015/10/16-asselborn-avramopoulos-lesbos/index.html.

51. Interview with the Chios general police director, Chios, March 1, 2016.

52. "Frontex Accepts Greece's Request for Rapid Border Intervention Teams," Frontex, accessed December 8, 2017, http://frontex.europa.eu/news/frontex-accepts-greece -s-request-for-rapid-border-intervention-teams-amcPjC.

53. Interview with the mayor of the municipality of Chios, Chios, March 7, 2016.

54. Interview with the mayor of the municipality of Chios, Chios, March 7, 2016.

55. Interview with the Chios deputy head of the coast guard, Chios, March 2, 2016. Some days later, he was upgraded to the head of the Chios coast guard. At that time, he acted as HCG head.

56. Interview with the mayor, municipality of Chios, Chios, March 7, 2016.

57. Interview with the mayor, municipality of Lesbos, Lesbos, February 26, 2016.

58. Interview with the mayor, municipality of Lesbos, Lesbos, February 26, 2016.

59. As Karolina Follis argues in "Vision and Transterritory: The Borders of Europe," *Science, Technology, and Human Values* 42, no. 6 (2017): 1016, "transterritorial vision such as that produced by Eurosur likewise may offer the illusion of transparent surveillance at what one document describes as a 'nonnegligible' distance (European Commission JRC 2015). And yet, the picture it delivers is always already compromised by the oligoptic characteristics of its own infrastructure and by the culturally and politically mediated preconceptions of its embedded actors."

60. See Ian Chambers, *Mediterranean Crossings: The Politics of an Interrupted Modernity* (Durham, NC/London: Duke University Press, 2008); and Franco Cassano, *Southern*

Thought and Other Essays on the Mediterranean (New York: Fordham University Press, 2012).

61. As Kerem Öktem states in "The Ambivalent Sea: Regionalizing the Mediterranean Differently," *Mediterranean Frontiers. Borders, Conflict and Memory in a Transnational World* (London/New York: Tauris Academic Studies, 2010), "one of the tensions that run through his [Braudel's] epic work is the conflict between the claim to Mediterranean unity, and the very fact of the region's political and religious division." As Chambers, in *Mediterranean Crossings*, 49, puts it: "The space of the Mediterranean, both as sea and combinatory territory, remains elusive: a perpetual interrogation. The sea is not something to possess . . . If there is a unity in the Mediterranean, it is perhaps a hidden, critical 'unity' where the sea itself, as the site of dispersion and drift, exposes the fragility of inherited configurations."

62. Edward Said, *Orientalism: Western Conceptions of the Orient* (London: Penguin, 1977).

63. Luigi Cazzato, "An Archaeology of the Verticalist Mediterranean: From Bridges to Walls," *Mediterranean Review* 5, no. 2 (2012): 28.

64. Fernand Braudel, *The Mediterranean and the Mediterranean World in the Age of Philip II* (Berkeley: University of California Press, 1996). See also Barry Ryan, "Security Spheres: A Phenomenology of Maritime Spatial Practices," *Security Dialogue* 46, no. 6 (2015): 569.

65. Geoffrey A. Boyce, "The Rugged Border: Surveillance, Policing and the Dynamic Materiality of the US/Mexico Frontier," *Environment and Planning D: Society and Space* 34, no. 2 (2016); and Squire (2014; 2017).

66. Boyce, "The Rugged Border," 256.

67. Boyce, "The Rugged Border," 257.

68. Vicky Squire, "Desert 'Trash': Posthumanism, Border Struggles, and Humanitarian Politics," *Political Geography* 39 (2014): 12.

69. Didier Bigo, "Death in the Mediterranean Sea: The Results of the Three Fields of Action of European Union Border Controls," *The Irregularization of Migration in Contemporary Europe*, ed. Jansen Yolande, Robin Celikates and Joost de Bloois (London/New York: Rowman and Littlefield, 2015).

70. Bigo, "Death in the Mediterranean Sea," 59.

71. This development accords with what Barry Ryan, in "Security Spheres: A Phenomenology of Maritime Spatial Practices," *Security Dialogue* 46, no. 6 (2015): 579–580, describes as the securitization of maritime borders to design three-dimensional environments.

72. Charles Heller and Lorenzo Pezzani, "Liquid Traces. Investigating the Deaths of Migrants at the EU's Maritime Frontier," in *The Borders of "Europe." Autonomy of*

Migration, Tactics of Bordering, ed. Nicholas De Genova (Durham, NC/London: Duke University Press, 2017), 103.

73. Heller and Pezzani, "Liquid Traces," 103.

74. Heller and Pezzani, "Liquid Traces," 103.

75. Cf. William Walters, "Migration, Vehicles, and Politics: Three Theses on Viapolitics," *European Journal of Social Theory* 18, no. 4 (2015): 469–488.

76. Bruno Latour, *Reassembling the Social: An Introduction to Actor-Network-Theory* (Oxford: Oxford University Press, 2005), 181.

77. Gerard de Vries, *Bruno Latour* (Cambridge, UK: Polity Press, 2016), 96.

Chapter 6

1. Katerina Rozakou, "Socialities of Solidarity: Revisiting the Gift Taboo in Times of Crises," *Social Anthropology* 24, no. 2 (2016): 196.

2. The concept of the "humanitarian border" is coined by William Walters in "Foucault and Frontiers: Notes on the Birth of the Humanitarian Border," in *Governmentality: Current Issues and Future Challenges*, ed. Ulrich Bröckling, Suzanne Krasmann, and Thomas Lemke (New York: Routledge, 2011).

3. "Human security" is an important international approach and concept in security studies. See, for instance, https://www.un.org/humansecurity/what-is-human -security/.

4. Chris Rumford. *Citizens and Borderwork in Contemporary Europe* (London: Routledge, 2008) and Polly Pallister-Wilkins, "The Humanitarian Politics of European Border Policing: Frontex and Border Police in Evros," *International Political Sociology* 9, no. 1 (2015): 53–69.

5. Didier Fassin, *Humanitarian Reason: A Moral History of the Present* (Berkeley: University of California Press, 2012).

6. Fassin, *Humanitarian Reason*, 135–138.

7. Didier Fassin, "Humanitarianism as a Politics of Life," *Public Culture* 19, no. 3 (2007): 519.

8. Michael Barnett, *Empire of Humanity: A History of Humanitarianism* (Ithaca, NY/ London: Cornell University Press, 2011).

9. As in chapter 5, the interviews were organized, conducted, transcribed, and translated by Ermioni Frezouli. For a more specific analysis of the movability of the humanitarian border based on this research, see Huub Dijstelbloem and Lieke van der Veer, "The Multiple Movements of the Humanitarian Border: The Portable Provision of Care and Control at the Aegean Islands," *Journal of Borderlands Studies* (2019): 1–19.

10. Bernd Kasparek, "Routes, Corridors, and Spaces of Exception: Governing Migration and Europe," *Near Futures Online* 1 "Europe at a Crossroads," March 2016, http://nearfuturesonline.org/routes-corridors-and-spaces-of-exception-governing-migration-and-europe/.

11. See Alison Mountz, "The Enforcement Archipelago: Detention, Haunting, and Asylum on Islands," *Political Geography* 30 (2011): 118–128; and Allison Mountz, "Political Geography II: Islands and Archipelagos," *Progress in Human Geography* 39, no. 5 (2014): 636–646.

12. The aim is to study the islands as what Anne-Laure Amilhat Szary and Frédéric Giraut (2015, 3), in their introduction to *Borderities and Politics of Contemporary Mobile Borders* (Basingstoke, UK: Palgrave Macmillan, 2015), call a "laboratory for studying socio-spatial relations" that stages "the multiple rules and experiences of what a border can be."

13. My visit to Moria, September 10, 2014, with Rogier van Reekum and Ermioni Frezouli.

14. Walters, "Foucault and Frontiers," 146.

15. See Reece Jones, Corey Johnson, Wendy Brown, Gabriel Popescu, Polly Pallister-Wilkins, Alison Mountz, and Emily Gilbert, "Interventions on the State of Sovereignty at the Border," *Political Geography* 59 (2017): 6. As Chris Rumford, in *Citizens and Borderwork in Contemporary Europe* (London: Routledge, 2008), 5, points out, "people, not just states, engage in bordering activities."

16. Jones et al., "Interventions on the State of Sovereignty at the Border," 6.

17. Joel Hernandez, "Refugee Flows to Lesvos: Evolution of a Humanitarian Response," *Migration Policy Institute*, January 29, 2016, https://www.migrationpolicy.org/article/refugee-flows-lesvos-evolution-humanitarian-response.

18. Polly Pallister-Wilkins, "Humanitarian Borderwork: Actors, Spaces, Categories," in Jones et al., "Interventions on the State of Sovereignty at the Border," 9.

19. See Dijstelbloem and van der Veer, "The Multiple Movements of the Humanitarian Border." See also Maribel Casas-Cortes, Sebastian Cobarrubias, Nicholas De Genova, Glenda Garelli, Giorgio Grappi, Charles Heller, Sabine Hess et al., "New Keywords: Migration and Borders," *Cultural Studies* 29, no. 1 (2015): 55–87; Peter Nyers, "Moving Borders: The Politics of Dirt," *Radical Philosophy* 174 (July/August 2012); Walters, "Foucault and Frontiers."

20. Visit PIKPA, Lesbos, September 10 2014, with Rogier van Reekum and Ermioni Frezouli.

21. Petra is located 5 kilometers north of Molyvos, and it's a very touristy area as well.

22. Interview with a member of PIKPA, Lesbos, February 24, 2016.

23. Nyers, "Moving Borders," 174.

24. Interview with the commander of the Lesvian coast guard, Lesbos, February 25, 2016.

25. The analyses of the situations at Lesbos and Chios in these sections and the interpretations of the interviews were conducted in cooperation with Lieke van der Veer. The analysis is partly based on Dijstelbloem and van der Veer, "The Multiple Movements of the Humanitarian Border."

26. Evthymios Papataxiarchis, in "Being 'There': At the Front Line of the 'European Refugee Crisis'—Part 1," *Anthropology Today* 32, no. 2 (2016): 5, observes that "everyone and everything else goes where the refugees go."

27. Interview with a freelance journalist who worked on Chios for an extended period of time, Chios, May 8, 2016.

28. Annemarie Mol, "Actor-Network Theory: Sensitive Terms and Enduring Tensions," *Kölner Zeitschrift für Soziologie und Sozialpsychologie* 50, no. 1 (2010): 259.

29. Interview with the mayor of the municipality of Chios, Chios, March 7, 2016.

30. Interview with a volunteer working in the fishing port of Agia Ermioni, Chios, March 10, 2016.

31. Interview with a volunteer related to Chios Solidarity, Chios, May 27, 2016.

32. Lathra, accessed May 28, 2020, http://www.lathra.gr/.

33. Interview with a member of Lathra, Chios, May 19, 2016.

34. Norwegian Refugee Council, accessed May 28, 2020, https://www.nrc.no/who -we-are/about-us/.

35. Interview with the area manager of the Norwegian Refugee Council in Greece, Chios, May 12, 2016.

36. Interview with the area manager of the Norwegian Refugee Council in Greece, Chios, May 12, 2016.

37. Interview with a consultant who worked for the Ministry of Migration, Athens, February 12, 2016.

38. Didier Bigo, "Security and Immigration: Toward a Critique of the Governmentality of Unease," *Alternatives* 27, no. 1 (2002): 63–92.

39. Paolo Cuttitta, "Repoliticization Through Search and Rescue? Humanitarian NGOs and Migration Management in the Central Mediterranean," *Geopolitics* 23, no. 3 (2017): 632–660.

40. Interview with the area manager of the Norwegian Refugee Council in Greece, Chios, May 12, 2016.

41. Interview with the former deputy head of the Chian coast guard, Chios, March 2, 2016.

42. Interview with a member of Lathra, Chios, May 19, 2016.

43. "MSF No Longer Take Funds from EU Member States and Institutions," *MSF*, June 16, 2016 (updated December 8, 2016), https://www.msf.org.uk/article/msf-no -longer-take-funds-eu-member-states-and-institutions.

44. Interview with a volunteer working for Solidarity Kitchen, Chios, May 17, 2016.

45. "Humanitarian practice, like the border itself, is influenced by the settings and the types of work or practices carried out, and as such is always in a state of becoming," as Pallister-Wilkins in, "Humanitarian Borderwork," 20, argues.

46. Walters, "Foucault and Frontiers," 148.

47. Interview with a member of Starfish Foundation, Molyvos, Lesbos, February 24, 2016.

48. Interview with the mayor of the municipality of Chios, Chios, March 7, 2016.

49. "Who We Are," *Doctors of the World*, accessed May 28, 2020, https://www .doctorsoftheworld.org.uk/about-us.

50. Interview with an employee of Médecins du Monde, Chios, May 4, 2016.

51. Interview with a member of Starfish Foundation, Molyvos, Lesbos, February 24, 2016.

52. Interview with a consultant who worked for the Ministry of Migration, Athens, February 12, 2016.

53. Interview with the mayor of the municipality of Chios, Chios, March 7, 2016.

54. Interview with a member of Starfish Foundation, Molyvos, Lesbos, February 24, 2016.

55. See Katja Lindskov Jacobsen, "On Humanitarian Refugee Biometrics and New Forms of Intervention," *Journal of Intervention and Statebuilding* 11, no. 4 (2017): 529; and Benjamin Meiches, "Non-human Humanitarians," *Review of International Studies* 45, no. 1 (2019): 1.

56. Interview with the commander of the Lesvian coast guard, Lesbos, February 25, 2016.

57. Interview with a volunteer working for Solidarity Kitchen, Chios, May 17, 2016.

58. Arjun Appadurai, in *The Future as Cultural Fact. Essays on the Global Condition* (New York: Verso, 2013), 64, argues that it is crucial to understand the relationship "between the forms of circulation and the circulation of forms."

59. Interview with an employee of Médecins du Monde, Chios, May 4, 2016.

60. Interview with the mayor of the municipality of Chios, Chios, March 7, 2016.

61. Interview with the deputy head of the Regional Authority of the North Aegean, Chios, March 2, 2016.

62. Interview with the deputy head of administrative and financial services of the Regional Authority of North Aegean, Lesbos, February 26, 2016.

63. Interview with the deputy head of administrative and financial services of the Regional Authority of North Aegean, Lesbos, February 26, 2016.

64. Interview with the general police director, Chios, March 1, 2016.

65. Interview with a member of Lathra, Chios, May 19, 2016.

66. Interview with the area manager of the Norwegian Refugee Council in Greece, Chios, May 12, 2016. As Pallister-Wilkins, "Humanitarian Borderwork," 8, points out, "the emergent and ephemeral humanitarian borderscape is structured and conditioned by the im/mobility of migrants as they are channeled through 'corridors' and 'narrow bands' structured by border controls and transport infrastructures."

67. Brigitta Kuster and Vasilis Tsianos, "Erase Them: Eurodac and Digital Deportability," *Transversal Texts*, no. 2 (2013): http://eipcp.net/transversal/0313/kuster-tsianos/en.

68. Vasilis Galis, Spyros Tzokas, and Aristotle Tympas, "Bodies Folded in Migrant Crypts: Dis/Ability and the Material Culture of Border-Crossing," *Societies* 6, no. 10 (2016): 1–11.

69. Interview with an employee of Médecins du Monde, Chios, May 4, 2016.

70. Interview with the former deputy head of the Chian coast guard, Chios, March 2, 2016.

71. Papataxiarchis, "Being 'There,'" 6–7.

72. Interview with a member of Starfish Foundation, Molyvos, Lesbos, February 24, 2016.

73. Interview with a freelance journalist who worked on Chios for an extended period of time, Chios, May 8, 2016.

74. As Pallister-Wilkins points out in "Humanitarian Borderwork," 7: "Humanitarian borderwork is very diverse in terms of actors and border spaces because of its contingent relationship with mobility. Fluctuating assemblages of actors with divergent socio-political objectives undertake humanitarian borderwork. These actors share humanitarian sensibilities but perform acts of rescue and caregiving for reasons that cannot be considered 'wholly' humanitarian ... Humanitarian borderwork is therefore indicative of the instrumental and normative logics present in much humanitarian work, where pragmatic security concerns intersect with affective concerns for people's wellbeing and dignity."

75. Interview with a member of Lathra, Chios, May 19, 2016.

76. Interview with the mayor of the municipality of Chios, Chios, March 7, 2016.

77. Interview with the director of Politis Chios, a journal and news blog, Chios, May 10, 2016.

78. Interview with a freelance journalist who worked on Chios for an extended period of time, Chios, May 8, 2016.

79. Interview with an employee of Médecins du Monde, Chios, May 4, 2016.

80. Interview with a member of Starfish Foundation, Molyvos, Lesbos, February 24, 2016.

81. Interview with a freelance journalist who worked on Chios for an extended period of time, Chios, May 8, 2016.

82. Martina Tazzioli, "Containment through Mobility: Migrants' Spatial Disobediences and the Reshaping of Control through the Hotspot System," *Journal of Ethnic and Migration Studies* 44, no. 16 (2018): 2764–2779.

83. Nyers, "Moving Borders."

84. See Lindskov Jacobsen, "On Humanitarian Refugee Biometrics and New Forms of Intervention," 529–551; and Meiches, "Non-human Humanitarians," 1–19.

Chapter 7

1. "The List: The 34,361 Men, Women and Children Who Perished Trying to Reach Europe," *The Guardian*, June 20, 2018, https://www.theguardian.com/world/2018 /jun/20/the-list-34361-men-women-and-children-who-perished-trying-to-reach -europe-world-refugee-day.

2. "People for Sale," CNN, October 2017, http://edition.cnn.com/specials/africa /libya-slave-auctions.

3. "Greece: Frontier of Hope and Fear," Amnesty International, April 29, 2014, https://www.amnesty.org/en/documents/EUR25/004/2014/en/.

4. John Torpey, *The Invention of the Passport: Surveillance, Citizenship and the State* (Cambridge: Cambridge University Press, 1998).

5. James C. Scott, *Seeing Like a State: How Certain Schemes to Improve the Human Condition Have Failed* (New Haven, CT/London: Yale University Press, 1998), 9.

6. See Marieke de Goede, Anna Leander, and Gavin Sullivan, "Introduction: The Politics of the List," *Environment and Planning D: Society and Space* 32, no. 1 (2016): 4–5.

7. According to Anna Leander, "The Politics of Whitelisting: Regulatory Work and Topologies in Commercial Security," *Environment and Planning D: Society and Space*

34, no. 1 (2016): 48–66, lists have agency: "they can make things happen and can therefore be held co-responsible for political developments."

8. Umberto Eco would probably have agreed: "Faced with something that is immensely large, or unknown, of which we still do not know enough or of which we shall never know, the author proposes a list as specimen, or indication, leaving the reader to imagine the rest" (Umberto Eco, *The Infinity of Lists: From Homer to Joyce* (London: MacLehose Press, 2009), quoted in De Goede et al., 7.

9. The discussion of the politics of the list thus can be continued with what Stäheli, in "Indexing—The Politics of Invisibility," 15, calls their "politics of invisibility": "list-making is not only a problem of selection, but it is necessarily a transformative and performative practice: it produces the items which the list will comprise." Her argument thus speaks to Geoffrey C. Bowker's and Susan Leigh Star's observation in *Sorting Things Out: Classification and Its Consequences* (Cambridge, MA: MIT Press, 1999), a groundbreaking work on the nature of classifications and their consequences, that "the material culture of bureaucracy and empire is not found in pomp and circumstance, nor even in the first instance of the point of a gun, but rather at the point of a list." While Stäheli mainly meant lists made by state agents, the same would seem to hold for lists made by nonstate actors. The making of lists also recalls the notion of "agnotology," introduced by Robert Proctor in "Agnotology: A Missing Term to Describe the Cultural Production of Ignorance (and Its Study)," in *Agnotology: The Making and Unmaking of Ignorance*, ed. Robert Proctor and Londa Schiebinger (Stanford, CA: Stanford University Press, 2008), 1–37—that is, the study of ignorance and the publication of inaccurate or misleading data.

10. Tamara Last, Giorgia Mirto, Orçun Ulusoy, Ignacio Urquijo, Joke Harte, Nefeli Bami, Marta Pérez Pérez, et al., "Deaths at the Borders Database: Evidence of Deceased Migrants' Bodies Found along the Southern External Borders of the European Union," *Journal of Ethnic and Migration Studies* 43, no. 5 (2017): 693–712.

11. Last et al., "Deaths at the Borders Database, 694.

12. Tamara Last and Thomas Spijkerboer, "Tracking Deaths in the Mediterranean," in *Fatal Journeys. Tracking Lives Lost during Migration*, ed. Tara Brian and Frank Laczko (Geneva, Switzerland: IOM, 2014).

13. Last et al., "Deaths at the Borders Database, 709.

14. Donna Haraway, *Feminist Studies* 14, no. 3 (1988): 581.

15. The following discussion elaborates on the analysis in Huub Dijstelbloem, "Migration Tracking Is a Mess," *Nature* 543, no. 2 (2017): 32–34.

16. "Operations Portal, Refugee Situations," UNHCR, accessed May 31, 2020, https://data2.unhcr.org/en/situations.

17. IOM, *World Migration Report 2018* (Geneva, Switzerland: IOM, 2017), 160.

18. "WatchTheMed Alarm Phone Reports," *Watch the Med*, accessed May 31, 2020, http://www.watchthemed.net/.

19. "About," Watch the Med Alarm Phone, accessed May 31, 2020, https://alarm phone.org/en/about/.

20. See Dijstelbloem, "Migration Tracking Is a Mess."

21. "Our Story," Satellite Sentinel Project, accessed May 31, 2020, http://www.satsen tinel.org/.

22. Marouf Hasian, Jr., *Forensic Rhetorics and Satellite Surveillance: The Visualization of War Crimes and Human Rights Violations* (Lanham, MD: Lexington Books, 2016).

23. IOM, *World Migration Report 2018*, 162.

24. IOM, *World Migration Report 2018*, 160–162.

25. The term was coined by Nicholas De Genova, "Spectacles of Migrant 'Illegality': The Scene of Exclusion, the Obscene of Inclusion," *Ethnic and Racial Studies* 36, no. 7 (2013): 1180. See for follow-ups on this topic also for instance Nicholas De Genova, "Introduction: The Borders of 'Europe' and the European Question," in *The Borders of "Europe": Autonomy of Migration, Tactics of Bordering*, ed. Nicholas De Genova (Durham, NC/London: Duke University Press, 2017), 3–9.

26. Dijstelbloem, "Migration Tracking Is a Mess," 33.

27. See Huub Dijstelbloem, Rogier van Reekum, and Willem Schinkel, "Surveillance at Sea: The Transactional Politics of Border Control in the Aegean," *Security Dialogue* 48, no. 3 (2017): 224–240.

28. "Refugee Narratives: Case Farmakonisi or The Justice of the Water," *Critical Stages*, no. 14 (2016), accessed May 31, 2020, http://www.critical-stages.org/14/refugee -narratives-case-farmakonisi-or-the-justice-of-the-water-by-anestis-azas/.

29. "Tools and Reports," Amnesty International, accessed May 31, 2020, www .amnestyusa.org/research/science-for-human-rights.

30. *Push Back Map*, accessed May 31, 2020, https://pushbackmap.org/.

31. "Welcome to eyeWitness," *eyeWitness*, accessed May 31, 2020, www.eyewitness project.org/welcome.

32. "Border Crossing Observatory," accessed November 12, 2020, https://www.mon ash.edu/arts/border-crossing-observatory/about-us#:~:text=The%20Border%20 Crossing%20Observatory%20is,edge%20research%20on%20border%20crossings.

33. "Arizona OpenGIS Initiative for Deceased Migrants," *Humane Borders*, accessed May 31, 2020, http://humaneborders.info/.

34. "The Human and Financial Cost of 15 Years of Fortress Europe," *The Migrants' Files*, accessed May 31, 2020, http://www.themigrantsfiles.com/.

35. "Rethinking Journalism with Data," J++, accessed May 31, 2020, http://jplusplus .org/en/.

36. IOM, *World Migration Report 2018*, 157, 158.

37. Koen Leurs, *Digital Passages: Migrant Youth 2.0. Diaspora, Gender and Youth Cultural Intersections* (Amsterdam: Amsterdam University Press, 2015), 14.

38 Kevin Smets, "The Way Syrian Refugees in Turkey Use Media: Understanding 'Connected Refugees' through a Non-mediacentric and Local Approach," *Communications* 43, no. 1 (2018): 113.

39. Sandra Ponzanesi, "Digital Diaspora's: Postcoloniality, Media and Affect," *Interventions* (2020): 7–11.

40. "The Left-to-Die Boat," Forensic Architecture, accessed May 31, 2020, http:// www.forensic-architecture.org/case/left-die-boat.

41. Charles Heller and Lorenzo Pezzani, "Ebbing and Flowing: The EU's Shifting Practices of (Non-)Assistance and Bordering in a Time of Crisis," *Near Futures Online Issue 1 "Europe at a Crossroads,"* March 2016, accessed May 31, 2020, http://nearfuturesonline.org/ebbing-and-flowing-the-eus-shifting-practices-of-non -assistance-and-bordering-in-a-time-of-crisis/.

42. Ulrich Beck, *World at Risk* (Cambridge, UK: Polity, 2009), 27.

43. Heller and Pezzani, "Ebbing and Flowing."

44. According to Dennis Rodgers and Bruce O'Neill, in "Infrastructural Violence: Introduction to the Special Issue," *Ethnography* 13, no. 4 (2012): 403, the term 'infrastructural violence' "explicitly draws attention to the fact that the workings of infrastructure can be substantially deleterious." For Rodgers and O'Neill (2012, 404), infrastructures constitute "an often ignored material channel for what is regularly referred to as 'structural violence,' which Paul Farmer has defined as 'violence exerted systemically—that is, indirectly—by everyone who belongs to a certain social order.'" See Paul Farmer, "An Anthropology of Structural Violence," *Current Anthropology* 45, no. 3 (2004): 307. The notion of infrastructural violence can thus be read as the less state-oriented successor of what Michael Mann, in "The Autonomous Power of the State: Its Origins, Mechanisms and Results," *European Journal of Sociology* 25, no. 2 (1984): 113, called "infrastructural power"; that is, "the capacity of the state actually to penetrate civil society, and to implement logistically political decisions throughout the realm."

45. Donna Haraway, *Feminist Studies* 14, no. 3 (1988): 581.

46. Donna Haraway, *Feminist Studies*, 581.

47. Lorraine Daston and Peter Galison. *Objectivity* (Cambridge, MA: MIT Press, 2007).

48. Eyal Weizman, *Forensic Architecture: Violence at the Threshold of Detectability* (New York: Zone Books, 2017), 9.

49. See Louise Amoore, "Biometric Borders: Governing Mobilities in the War on Terror," *Political Geography* 25 (2006): 336–351; Louise Amoore, "Lines of Sight: On the Visualization of Unknown Futures," *Citizenship Studies* 13, no. 1 (2009): 17–30; Louise Amoore and Alexandra Hall, "Border Theatre: On the Arts of Security and Resistance," *Cultural Geographies* 17, no. 3 (2010): 299–319; and Louise Amoore and Alexandra Hall, "The Clown at the Gates of the Camp: Sovereignty, Resistance and the Figure of the Fool," *Security Dialogue* 44, no. 2 (2014): 93–110.

50. The notion is developed by the art historian Jonathan Crary, in *Techniques of the Observer: On Vision and Modernity in the 19th Century* (Cambridge, MA: MIT Press, 1992), in which he describes the historical relationship between ideas about vision and subjectivity. As he explains in *Suspensions of Perception: Attention, Spectacle, and Modern Culture* (Cambridge, MA: MIT Press, 1999), 3 "vision is only one part of a body capable of evading institutional capture and of inventing new forms, affects, and intensities."

51. Louise Amoore, "Lines of Sight," 19, 22.

52. Luc Boltanski, *Distant Suffering: Morality, Media and Politics* (Cambridge: Cambridge University Press, 1999).

53. Crary, *Techniques of the Observer*, 9–11.

54. In many ways, these arguments speak to Foucault's oeuvre, especially his thesis in *Discipline and Punishment* (New York: Random House, 1975). According to Crary, Foucault, in emphasizing surveillance, pays little attention to the development of perception itself (see Crary, *Techniques of the Observer*, 19).

55. The argument in this section and the analysis of the "detective" is based in part on Huub Dijstelbloem, Chapter 7 "Seeing Like a Detective" [Kijken als een detective] of Huub Dijstelbloem, *The House of Argus: The Watchful Eye in Democracy* [*Het Huis van Argus: De Wakende Blik in de Democratie*] (Amsterdam: Boom, 2016), 114–131, but this discussion revises and updates the material in the book.

56. Bruno Latour, *Aramis or the Love of Technology* (Cambridge, MA: Harvard University Press, 1996).

57. Although it is dangerous to talk of periods that are strictly demarcated and clearly distinguished, it is still possible to distinguish three phases in the development of the detective novel. The first phase encompasses the Victorian detective novel, set in nineteenth-century Paris and London, which we know from Edgar Allan Poe and Sir Arthur Conan Doyle. The second covers the rise of the hardboiled

detective in the works of Dashiell Hammett and Raymond Chandler, in San Francisco and Los Angeles in early twentieth-century California. The third phase relates to what is called the postwar metaphysical or postmodern thriller, such as in the works of Paul Auster. Before that time, there were riddles to be found in the literature, exciting stories and characters deciphering deeds or misdeeds, such as in Voltaire's *Zadig*, but these can hardly be classified as detective literature. See, for instance, Josef Hoffmann, *Philosophies of Crime Fiction* (Harpenden/Hertfordshire, UK: No Exit Press, 2013); Stephen Knight. *Crime Fiction Since 1800. Detection, Death, Diversity* (Basingstoke, UK: Palgrave MacMillan, 2010); Martin Priestman (ed.), *The Cambridge Companion to Crime Fiction* (Cambridge: Cambridge University Press, 2010); John Scaggs, *Crime Fiction* (London/New York: Routledge, 2005).

58. It is striking that Dostoyevsky's *Crime and Punishment* is not mentioned in any of the historical overviews that I have consulted. That would certainly enrich the description of the genre, because while the novel involves a murder, a perpetrator, and an investigator, it also explores above all the idea that it is the inner detection work by the main character and the search within his conscience—the question of whether he has a conscience at all—that ramps up the tension and makes the investigative work so intimate that the reader almost becomes an accomplice.

59. Luc Boltanski, *Mysteries and Conspiracies: Detective Novels, Spy Novels and the Making of Modern Societies.* (Oxford, UK: Wiley-Blackwell, 2014), 112: "The detective is the state in a normalised state of emergency."

60. Foucault, *Discipline and Punishment.*

61. His personal favorite was no. 50: *Deadlier than the Male*, the only novel by James Gunn, which was published in 1943.

62. Scott Dimovitz, "Public Personae and the Private I. De-compositional Ontology in Paul Auster's *The New York Trilogy*," *Modern Fiction Studies* 52, no. 3 (2006): 613–633.

63. Hasian, *Forensic Rhetorics and Satellite Surveillance.*

64. Paul Auster, *The New York Trilogy* (New York: Penguin, 1987), 8.

65. Dimovitz, "Public Personae and the Private I," 620.

66. "Crossing the Mediterranean Sea by Boat," Crossing the Med Map, accessed June 1, 2020, https://crossing-the-med-map.warwick.ac.uk/.

67. *Migration Trail*, accessed June 1, 2020, https://www.migrationtrail.com/.

68. This corresponds with what Jacques Rancière argues in *The Politics of Aesthetics: The Distribution of the Sensible* (London and New York: Continuum, 2004).

69. Jacques Rancière, *Dissensus: On Politics and Aesthetics* (London: Bloomsbury, 2010), 70.

70. See Nicholas Mirzoeff, *The Right to Look: A Counterhistory of Visuality* (Durham, NC/London: Duke University Press, 2011), which specifically studies relationships between the visible and the invisible. The right to look, Mirzoeff argues, is the right to escape from the totalizing effect of the visual—to not only be represented according to others, but to look and develop one's own perspective.

71. As Alison Mountz argues in *Seeking Asylum: Human Smuggling and Bureaucracy at the Border* (Minneapolis: University of Minnesota Press, 2010), 23, "visibility proves crucial to understanding how states respond to migrants . . . during highly publicized, visible, visual, and seemingly exceptional crises along their borders."

72. Jeffrey Green, *The Eyes of the People: Democracy in an Age of Spectatorship* (Oxford: Oxford University Press, 2011), 9.

73. Nadia Urbinati, *Democracy Disfigured: Opinion, Truth, and the People* (Cambridge, MA: Harvard University Press, 2014).

74. Green in *The Eyes of the People*, discusses how the Caesarian element can then enter into politics (he discusses Shakespeare's Roman tragedies of Julius Caesar, Coriolanus, and Antonius and Cleopatra). As Weber argued, the Caesarian style can imbue leadership with charisma that the bureaucracy and procedural and deliberative institutions cannot, at the cost of denying individuality to citizens.

75. In *Eyes* (Londen: Bloomsburry, 2015), Michel Serres examines the links between seeing, thinking, and knowing, and, in contrast to Plato, he invites the reader to remain in the semidark underground of the cave rather than seeing the true form of reality in the full light of the Sun. After all, this is where the search must start, in the twilight and the shadows where nothing is completely clear. For this reason, Serres views the philosopher as a kind of detective.

76. Nick Vaughan-Williams concludes, in "Borderwork beyond Inside/Outside? Frontex, the Citizen-Detective and the War on Terror," *Space and Polity* 12, no. 1 (2008): 76–77, "The figure of the 'citizen-detective' is likely to mobilize a vigilant subject constantly on the look-out for suspicious behavior not only in civic places but also rather closer to home. The promotion of this form of surveillance constitutes a form of generalized borderwork whereby, again, the borders of sovereign community are (re)produced not only at the edge of territories but throughout society at large."

77. "The gaze" is a translation of Foucault's account of the *le regard*, which he developed in *The Birth of the Clinic*. Andrew Barry emphasizes in *Political Machines: Governing a Technological Society* (London: Athlone Press, 2001), 56, the original term was meant "to refer to the process of making things visible rather than the act of looking as such."

78. Urbinati, *Democracy Disfigured*.

Chapter 8

1. Cf. the "pneumatic parliament" that was conceptually developed by Peter Sloterdijk and Gesa Mueller von der Hagen, see "Instant Democracy: The Pneumatic Parliament," accessed November 16, 2020, https://iffr.com/en/2006/films/instant-democracy-the -pneumatic-parliament.

2. See also the distinction made by Étienne Balibar in *Brexit, the Anti-Grexit*, accessed November 16, 2020, https://www.versobooks.com/blogs/2735-brexit-the-anti-grexit, between "internal exclusion" and "external inclusion." According to Balibar, Greece during its euro crisis was controlled by numerous policies in an internal exclusion process. The opposite happened with the Brexit crisis—negotiations with the UK were conducted in such a way that ties were kept as close as possible without being in the European Union through a process of external inclusion.

3. As Polly Pallister-Wilkins argues regarding security barriers in "How Walls Do Work: Security Barriers as Devices of Interruption and Data Capture," *Security Dialogue* 47, no. 2 (2016): 157, "What was a fence made up of wire or a wall made up of con- crete comes to enact an interruption in circulation, creating as it does so the time and space for the intervention of other devices and practices. As a device of interruption, security barriers work to configure a space in some way, creating a barrier using a range of materials that cannot be easily penetrated or moved through, restricting the movement of people and goods to materially deliberate spaces such as gates, crossing points and checkpoints that form an integral part of the device itself."

4. See Geoffrey Bowker and Susan Leigh Star, *Sorting Things Out. Classification and Its Consequences* (Cambridge, MA: MIT Press, 1999); David Lyon, *Surveillance after September 11* (Cambridge, UK: Polity Press, 2003); Huub Dijstelbloem and Dennis Broeders, "Border Surveillance, Mobility Management and the Shaping of Non- publics in Europe," *European Journal of Social Theory* 18, no. 1 (2014): 21–38.

5. The term "hidden integration" of Europe was coined by Thomas Misa and Johan Schot in "Introduction," *History and Technology* 21, no. 1 (2005): 1–19.

6. The term "boundary project" refers to the notion of "boundary objects," as devel- oped by Bowker and Leigh Star, *Sorting Things Out.*

7. Jo Guldi describes in *Roads to Power: Britain Invents the Infrastructure State* (Cam- bridge, MA: Harvard University Press, 2012) how British towns in the eighteenth and nineteenth centuries were connected by the construction of roads. It was not a fluid transition; all kinds of controversies arose as the new road connections implied new divisions and exclusions. Guldi illustrates the kind of government and bureaucratic force required to bring about this change as the rise of the "infrastructural state."

8. Peter Sloterdijk, *Not Saved: Essay after Heidegger* (Cambridge, UK: Polity Press, 2017), 90–95.

Coda

1. Stuart Blume, "Get Ready for the Global Fight over Vaccines," *New York Times*, April 30, 2020, SR3, https://www.nytimes.com/2020/04/30/opinion/sunday/coronavirus -vaccine-supply.html?searchResultPosition=1.

2. According to this code, "the reintroduction of border control at the internal borders must remain an exception and must respect the principle of proportionality. The scope and duration of such a temporary reintroduction of border control at the internal borders is limited in time and should be restricted to the bare minimum needed to respond to the threat in question. Reintroducing border control at the internal border should only ever be used as a measure of last resort," Regulation (EU) 2016/399 of the European Parliament and of the Council of March 9, 2016, on a Union Code on the Rules Governing the Movement of Persons across Borders (Schengen Borders Code), and "Temporary Reintroduction of Border Control," *European Commission*, accessed May 30, 2020, https://ec.europa.eu/home-affairs/what-we -do/policies/borders-and-visas/schengen/reintroduction-border-control_en.

3. Paola Tamma and Hannie Cokelaere, "Schengen Proves Hard to Reboot after System Meltdown," *Politico*, accessed June 1, 2020, https://www.politico.eu/article /schengen-proves-hard-to-reboot-after-system-meltdown/.

4. The following partly draws on Huub Dijstelbloem, "Bordering a Hybrid World: Infrastructural Isolation and the Governance of Human and Nonhuman Mobility," *Global Perspectives* 1, no. 1 (2020): 12789, https://doi.org/10.1525/gp.2020.12789.

5. See Louise Bengtsson, Stefan Borg, and Mark Rhinard, "Assembling European Health Security: Epidemic Intelligence and the Hunt for Cross-Border Health Threats," *Security Dialogue* 50, no. 2 (2019): 115–130. An early example of this development is the Monitoring Emerging Diseases (ProMed-Mail) Program, founded in 1994 by the Federation of American Scientists, which took advantage of the emerging online connectivity offered by the web. See Stephen Roberts and Stefan Elbe, "Catching the Flu: Syndromic Surveillance, Algorithmic Governmentality and Global Health Security," *Security Dialogue* 48, no. 1 (2017): 49. Other examples include the Global Public Health Intelligence Network (GPHIN), established by Canadian health authorities in 1997, and the Global Outbreak Alert and Response Network (GOARN), based on data from GPHIN and other sources. A final example is the Global Health Security Initiative (GHSI), a response by the G7 countries and Mexico to the 2001 anthrax attacks in the United States.

6. Examples include the ProMedProgram, the GPHIN program, and HealthMap.

7. See Bengtsson et al., "Assembling European Health Security," 122–123.

8. See Bengtsson et al., "Assembling European Health Security," 122–123.

9. Pierre Rosanvallon, *Counter-Democracy. Politics in an Age of Distrust* (Cambridge: Cambridge University Press, 2008), 30.

10. Here, I refer to James C. Scott, *Seeing Like a State. How Certain Schemes to Improve the Human Condition Have Failed* (New Haven, CT/London: Yale University Press, 1998).

11. See Charles Kenny, "Pandemics Close Borders—and Keep Them Closed," *Politico*, March 25, 2020, https://www.politico.com/news/magazine/2020/03/25/trump-corona virus-borders-history-plague-146788 ; and A. Wess Mitchell and Charles Ingrao, "Emperor Joseph's Solution to Coronavirus," *Wall Street Journal*, April 6, 2020, https://www.wsj.com/articles/emperor-josephs-solution-to-coronavirus-11586214561.

References

Agamben, Giorgio. *Homo Sacer. Sovereign Power, and Bare Life*. Stanford, CA: Stanford University Press, 1998.

Allen, John. *Topologies of Power. Beyond Territory and Network*. Oxford, UK/New York: Routledge, 2016.

Allen, Michael Thad, and Gabrielle Hecht. *Technologies of Power. Essays in Honor of Thomas Parke Hughes and Agatha Chipley Hughes*. Cambridge, MA: MIT Press, 2001.

Amilhat Szary, Laure, and Frédéric Giraut, eds. *Borderities and Politics of Contemporary Mobile Borders*. Basingstoke, UK: Palgrave Macmillan, 2015.

Amoore, Louise. "Biometric Borders: Governing Mobilities in the War on Terror." *Political Geography* 25 (2006): 336–351.

Amoore, Louise. "Lines of Sight: On the Visualization of Unknown Futures." *Citizenship Studies* 13, no. 1 (2009): 17–30.

Amoore, Louise. *The Politics of Possibility: Risk and Uncertainty Beyond Probability*. Durham, NC: Duke University Press, 2013.

Amoore, Louise. "Cloud Geographies: Computing, Data, Sovereignty." *Security Dialogue* 42, no. 1 (2018): 4–24.

Amoore, Louise, and Alexandra Hall. "Border Theatre: On the Arts of Security and Resistance." *Cultural Geographies* 17, no. 3 (2010): 299–319.

Amoore, Louise, and Alexandra Hall. "The Clown at the Gates of the Camp: Sovereignty, Resistance and the Figure of the Fool." *Security Dialogue* 44, no. 2 (2014): 93–110.

Anastasiadou, Irene. *Constructing Iron Europe: Transnationalism and Railways in the Interbellum*. Amsterdam: Amsterdam University Press, 2011.

Anderson, Benedict. *Imagined Communities*. New York: Verso, 1983.

Andersson, Ruben. *Illegality, Inc.: Clandestine Migration and the Business of Bordering Europe*. Oakland: University of California Press, 2014.

Andersson, Ruben. "Hardwiring the Frontier? The Politics of Security Technology in Europe's 'Fight against Illegal Migration.'" *Security Dialogue* 47, no. 1 (2016): 22–39.

Appadurai, Arjun. *The Future as Cultural Fact. Essays on the Global Condition*. New York: Verso, 2013.

Aradau, Claudia, and Tobias Blanke. "Politics of Prediction: Security and the Time/Space of Governmentality in the Age of Big Data." *European Journal of Social Theory* 20, no. 3 (2016): 373–391.

Aradau, Claudia, and Jef Huysmans. "Critical Methods in International Relations: The Politics of Tools, Devices and Acts." *European Journal of International Relations* 20, no 3 (2014): 596–619.

Augé, Mark. *Non-Places*. London/New York: Verso, 2008.

Auster, Paul. *The New York Trilogy*. New York: Penguin, 1987.

Balibar, Étienne. *We, the People of Europe? Reflections on Transnational Citizenship*. Princeton, NJ: Princeton University Press, 2004.

Balibar, Étienne. "Brexit, the Anti-Grexit." In *The Brexit Crisis: A Verso Report*. Edited by David Broder, 110–116. London: Verso, 2016.

Ballard, J. G. "Airports." *The Observer*, September 14, 1997.

Ballin, Ernst, Emina Ćerimović, Huub Dijstelbloem, and Mathieu Segers. *European Variations as a Key to Cooperation*. Cham, Switzerland: Springer Open, 2020.

Barnett, Michael. *Empire of Humanity: A History of Humanitarianism*. Ithaca, NY: Cornell University Press, 2011.

Barry, Andrew. *Material Politics: Disputes along the Pipeline*. Oxford, UK: Wiley-Blackwell, 2013.

Barry, Andrew. *Political Machines: Governing a Technological Society*. London: Athlone Press, 2001.

Barry, Andrew. "The Translation Zone: Between Actor-Network Theory and International Relations." *Millennium: Journal of International Studies* 41, no. 3 (2013): 413–429.

Basu, Natasha. *Is This Civil? Transnationalism, Migration and Feminism in Civil Disobedience*. PhD thesis, University of Amsterdam, 2019.

Beck, Ulrich. *World at Risk*. Cambridge, UK: Polity, 2009.

Bengtsson, Louise, Stefan Borg, and Mark Rhinard. "Assembling European Health Security: Epidemic Intelligence and the Hunt for Cross-Border Health Threats." *Security Dialogue* 50, no. 2 (2019): 115–130.

Besters, Michiel, and Frans Brom. "'Greedy' Information Technology: The Digitalization of the European Migration Policy." *European Journal of Migration and Law* 12, no. 4 (2010): 455–470.

Bialasiewicz, Luiza (Ed.). *Europe in the World: EU Geopolitics and the Making of European Space*. Surrey, UK: Ashgate Publishing Limited, 2011.

Bialasiewicz, Luiza. "Off-shoring and Out-sourcing the Borders of EUrope: Libya and EU Border Work in the Mediterranean." *Geopolitics* 17, no. 4 (2012): 843–866.

Bigo, Didier. "Death in the Mediterranean Sea: The Results of the Three Fields of Action of European Union Border Controls." In *The Irregularization of Migration in Contemporary Europe*. Edited by Jansen Yolande, Robin Celikates, and Joost de Bloois, 55–70. London/New York: Rowman and Littlefield, 2015.

Bigo, Didier. "Security and Immigration: Toward a Critique of the Governmentality of Unease." *Alternatives* 27, no. 1 (2002): 63–92.

Bijker, Wiebe, and John Law (Eds.). *Shaping Technology/Building Society: Studies in Sociotechnical Change*. Cambridge, MA: MIT Press, 1992.

Blume, Stuart. "Get Ready for the Global Fight over Vaccines." *The New York Times*, April 30, 2020.

Boltanski, Luc. *Distant Suffering: Morality, Media and Politics*. Cambridge: Cambridge University Press, 1999.

Boltanski, Luc. *Mysteries and Conspiracies: Detective Novels, Spy Novels and the Making of Modern Societies*. Oxford, UK: Wiley-Blackwell, 2014.

Boltanski, Luc, and Laurent Thévenot. *On Justification: Economies of Worth*. Princeton, NJ: Princeton University Press, 2006.

Bourne, Mike, Heather Johnson, and Debbie Lisle. "Laboratizing the Border: The Production, Translation and Anticipation of Security Technologies." *Security Dialogue* 46, no. 4 (2015): 307–325.

Bowker, Geoffrey C., and Susan Leigh Star. *Sorting Things Out: Classification and Its Consequences*. Cambridge, MA: MIT Press, 1999.

Boyce, Geoffrey A. "The Rugged Border: Surveillance, Policing and the Dynamic Materiality of the US/Mexico Frontier." *Environment and Planning D: Society and Space* 34, no. 2 (2016): 245–262.

Braudel, Fernand. *The Mediterranean and the Mediterranean World in the Age of Philip II*. Berkeley: University of California Press, 1996.

Bridle, James. "What They Don't Want You to See: The Hidden World of UK Deportation." *The Guardian*, January 27, 2015.

Broeders, Dennis. *Breaking Down Anonymity: Digital Surveillance of Irregular Migrants in Germany and the Netherlands*. Amsterdam: Amsterdam University Press, 2009.

Broeders, Dennis. "The New Digital Borders of Europe. EU Databases and the Surveillance of Irregular Migrants." *International Sociology* 22, no. 1 (2007): 71–92.

Broeders, Dennis, and Huub Dijstelbloem. "The Datafication of Mobility and Migration Management: The Mediating State and Its Consequences." In *Digitizing Identities: Doing Identity in a Networked World*. Edited by Irma van der Ploeg and Jason Pridmore, 242–260. New York and London: Routledge, 2015.

Brown, Mark. "Politicizing Science: Conceptions of Politics in Science and Technology Studies." *Social Studies of Science* 45, no 1 (2015): 3–30.

Brown, Wendy. *Walled States, Waning Sovereignty*. Cambridge, MA: MIT Press, 2011.

Callon, Michel, and Bruno Latour. "Unscrewing the Big Leviathan or How Do Actors Macrostructure Reality and How Sociologists Help Them to Do So." In *Advances in Social Theory and Methodology: Toward and Integration of Micro and Macro Sociologies*. Edited by K. Knorr Cetina and A. Cicourel, 277–303. London: Routledge and Kegan Paul, 1981.

Carrera, Sergio, Leonhard den Hertog, and Marco Stefan. "It Wasn't Me! The Luxembourg Court Orders on the EU-Turkey Refugee Deal." *CEPS Policy Insights* No. 2017/15, April 2017.

Casas-Cortes, Maribel, Sebastian Cobarrubias, Nicholas De Genova, Glenda Garelli, Giorgio Grappi, Charles Heller, Sabine Hess et al. "New Keywords: Migration and Borders." *Cultural Studies* 29, no. 1 (2015): 55–87.

Cassano, Franco. *Southern Thought and Other Essays on the Mediterranean*. New York: Fordham University Press, 2012.

Cazzato, Luigi. "An Archaeology of the Verticalist Mediterranean: From Bridges to Walls." *Mediterranean Review* 5, no. 2 (2012): 17–31.

Ceccorulli, Michela, and Sonia Lucarelli. "Securing the EU's Borders in the Twenty-First Century." In *EU Security Strategies: Extending the EU System of Security Governance*. 162–180. Edited by Spyros Economides and James Sperling. Abingdon, UK: Routledge, 2018.

Chambers, Ian. *Mediterranean Crossings: The Politics of an Interrupted Modernity*. Durham, NC/London: Duke University Press, 2008.

Ching, Francis D. K. *Form, Space, and Order*. New York: John Wiley and Sons, 1996.

Cisneros, Josue David. *Rhetorics of Borders, Citizenship and Latina/o Identity*. Tuscaloosa: University of Alabama Press, 2013.

Cole, Simon A. *Suspect Identities: A History of Fingerprinting*. Cambridge, MA: Harvard University Press, 2001.

Cooper, Anthony, Chris Perkins, and Chris Rumford. "The Vernacularization of Borders." In *Placing the Border in Everyday Life*. Edited by Reece Jones and Corey Johnson, 15–32. New York: Routledge, 2016.

Crary, Jonathan. *Suspensions of Perception: Attention, Spectacle, and Modern Culture*. Cambridge, MA: MIT Press, 1999.

Crary, Jonathan. *Techniques of the Observer: On Vision and Modernity in the 19th Century*. Cambridge, MA: MIT Press, 1992.

Cuttita, Paolo. "Pushing Migrants Back to Libya, Persecuting Rescue NGOs: The End of the Humanitarian Turn (Part I)," 2018a. Accessed January 4, 2019, https://www.law.ox.ac.uk/research-subject-groups/centre-criminology/centreborder-criminologies/blog/2018/04/pushing-migrants.

Cuttita, Paolo. "Pushing Migrants Back to Libya, Persecuting Rescue NGOs: The End of the Humanitarian Turn (Part II)," 2018b. Accessed January 4, 2019, https://www.law.ox.ac.uk/research-subject-groups/centre-criminology/centreborder-criminologies/blog/2018/04/pushing-0.

Cuttitta, Paolo. "Repoliticization through Search and Rescue? Humanitarian NGOs and Migration Management in the Central Mediterranean." *Geopolitics* 23, no. 3 (2017): 632–660.

Czaika, Mathias, and Hein de Haas. "The Globalization of Migration: Has the World Become More Migratory?" *International Migration Review* 48 (2014): 283–323.

Daston, Lorraine, and Peter Galison. *Objectivity*. Cambridge, MA: MIT Press, 2007.

De Cauter, Lieven. *De capsulaire beschaving. Over de stad in het tijdperk van de angst*. Rotterdam, Netherlands: NAI Uitgevers, 2009.

De Genova, Nicholas. "Introduction. The Borders of 'Europe' and the European Question." In *The Borders of "Europe." Autonomy of Migration, Tactics of Bordering*. Edited by Nicholas De Genova, 1–35. Durham, NC/London: Duke University Press, 2017.

De Genova, Nicholas. "Migrant 'Illegality' and Deportability in Everyday Life." *Annual Review of Anthropology* 31 (2002): 419–447.

De Genova, Nicholas. "The Queer Politics of Migration: Reflections on 'Illegality' and Incorrigibility." *Studies in Social Justice* 4, no. 2 (2010): 101–126.

De Genova, Nicholas. "Spectacles of Migrant 'Illegality': The Scene of Exclusion, the Obscene of Inclusion." *Ethnic and Racial Studies* 36, no. 7 (2013): 1180–1198.

De Goede, Marieke. "The Chain of Security." *Review of International Studies* 44, no. 1 (2018): 24–42.

De Goede, Marieke. "The Politics of Preemption and the War on Terror in Europe." *European Journal of International Relations* 14, no. 1 (2008): 161–185.

De Goede, Marieke, Anna Leander, and Gavin Sullivan. "Introduction: The Politics of the List." *Environment and Planning D: Society and Space* 32, no 1 (2016): 3–13.

De Goede, Marieke, Stephanie Simon, and Marijn Hoijtink. "Performing Preemption." *Security Dialogue* 45, no. 5 (2014): 411–422.

De Haas, Hein, Katharina Natter, and Simona Vezzoli. "Growing Restrictiveness or Changing Selection? The Nature and Evolution of Migration Policies." *International Migration Review* 52, no. 2 (June 2018): 324–367.

De Jong, Bart. *The Airport Assembled. Rethinking Planning and Policy Making of Amsterdam Airport Schiphol by Using the Actor-Network Theory.* PhD thesis, Utrecht University, 2012.

Den Heijer, Maarten. "Frontex and the Shifting Approaches to Boat Migration in the European Union: A Legal Analysis." In *Externalizing Migration Management. Europe, North America and the Spread of 'Remote Control' Practices.* Edited by Ruben Zaiotti, 53–71. London/New York: Routledge, 2016.

De Vries, Gerard. *Bruno Latour.* Cambridge, UK: Polity Press, 2016.

Dierikx, Mark, Johan Schot and Ad Vlot. "Van uithoek tot knooppunt: Schiphol." In *Techniek in Nederland in de Twintigste Eeuw, Deel V: Transport en Communicatie.* Edited by Johan Schot and Harry Lintsen. Zutphen: Walburg Pers, 2002.

Dijstelbloem, Huub. "Bordering a Hybrid World: Infrastructural Isolation and the Governance of Human and Nonhuman Mobility." *Global Perspectives* 1, no. 1 (2020): 12789.

Dijstelbloem, Huub. "Borders and the Contagious of Mediation." In *The SAGE Handbook of Media and Migration.* Edited by Kevin Smets, Koen Leurs, Myria Georghou, Saskia Witteborn, and Radhika Gajjala, 311–320. London: SAGE, 2020.

Dijstelbloem, Huub. *Het Huis van Argus: De Wakende Blik in de Democratie.* Amsterdam: Boom, 2016.

Dijstelbloem, Huub. "Mediating the Mediterranean: Surveillance and Counter-surveillance at the Southern Borders of Europe." In *The Irregularization of Migration in Contemporary Europe: Detention, Deportation, Drowning.* Edited by Yolande Jansen, Robin Celikates, and Joost de Bloois, 103–120. London: Rowman and Littlefield International, 2015.

Dijstelbloem, Huub. "Migration Tracking Is a Mess." *Nature* 543, no. 2 (2017): 32–34.

Dijstelbloem, Huub, and Dennis Broeders. "Border Surveillance, Mobility Management and the Shaping of Non-publics in Europe." *European Journal of Social Theory* 18, no. 1 (2014): 21–38.

Dijstelbloem, Huub, and Albert Meijer (Eds.). *Migration and the New Technological Borders of Europe.* Basingstoke, UK: Palgrave Macmillan, 2011.

Dijstelbloem, Huub, Rogier van Reekum, and Willem Schinkel. "Surveillance at Sea: The Transactional Politics of Border Control in the Aegean." *Security Dialogue* 48, no. 3 (2017): 224–240.

Dijstelbloem, Huub, and Lieke van der Veer. "The Multiple Movements of the Humanitarian Border: The Portable Provision of Care and Control at the Aegean Islands." *Journal of Borderlands Studies*, 2019. https:/doi.org 10.1080/08865655.2019.1567371.

Dijstelbloem, Huub, and William Walters. "Atmospheric Border Politics: The Morphology of Migration and Solidarity Practices in Europe," *Geopolitics*, 2019. https:/doi.org.10.1080/14650045.2019.1577826.

Dimovitz, Scott. "Public Personae and the Private I. De-compositional Ontology in Paul Auster's *The New York Trilogy.'" Modern Fiction Studies* 52, no. 3 (2006): 613–633.

Duez, Dennis, and Rocco Bellanova. "The Making (Sense) of EUROSUR: How to Control the Sea Borders?" In *EU Borders and Shifting Internal Security: Technology, Externalization, and Accountability*. Edited by Raphael Bossong and Helena Carrapico, 23–44. New York: Springer, 2016.

Dünnwald, Stephan. "Europe's Global Approach to Migration Management: Doing Border in Mali and Mauritania." In *Externalizing Migration Management: Europe, North America and the Spread of "Remote Control" Practices*. Edited by Ruben Zaiotti. London/New York: Routledge, 2016.

Dutch Safety Board. *Fire at the Detention Centre Schiphol Oost*. The Hague: Dutch Safety Board, 2006.

Eco, Umberto. *The Infinity of Lists. From Homer to Joyce*. London: MacLehose Press, 2009.

Edwards, Paul. "Infrastructure and Modernity: Scales of Force, Time, and Social Organization in the History of Sociotechnical Systems." In *Modernity and Technology*. Edited by Thomas Misa, Philip Brey, and Andrew Feenberg, 185–225. Cambridge, MA: MIT Press, 2003.

Edwards, Paul. "Meteorology as Infrastructural Globalism." *Osiris* 21: 229–250.

Edwards, Paul. *A Vast Machine: Computer Models, Climate Data, and the Politics of Global Warming*. Cambridge, MA: MIT Press, 2013.

Edwards, Paul, Matthew Mayernik, Archer Batcheller, Geoffrey Bowker, and Christine Borgman. "Science Friction: Data, Metadata, and Collaboration." *Social Studies of Science* 41, no. 5 (2011): 667–690.

Elam, Mark. "Living Dangerously with Bruno Latour in a Hybrid World." *Theory, Culture & Society* 16, no. 4 (1999): 1–24.

Elden, Stuart. *The Birth of Territory*. Chicago: University of Chicago Press, 2013.

Engelen, Ewald, Julie Froud, Sukhdev Johal, Angelo Salento, and Karel Williams. "How Cities Work: A Policy Agenda for the Grounded City." Cresc Working Paper Series, Working Paper 141, 2016.

Farmer, Paul. "An Anthropology of Structural Violence." *Current Anthropology* 45, no. 3 (2004): 305–325.

Fassin, Didier. "Humanitarianism as a Politics of Life." *Public Culture* 19, no. 3 (2007): 499–520.

Fassin, Didier. *Humanitarian Reason: A Moral History of the Present.* Berkely: University of California Press, 2012.

Feldman, Gregory. *The Migration Apparatus.* Stanford, CA: Stanford University Press, 2012.

FitzGerald, David Scott. *Refuge beyond Reach. How Rich Democracies Repel Asylum Seekers.* Oxford: Oxford University Press, 2019.

Follis, Karolina. "Vision and Transterritory: The Borders of Europe." *Science, Technology, and Human Values* 42, no. 6 (2017): 1003–1030.

Foucault, Michel. *Discipline and Punishment.* New York: Random House, 1975.

Frampton, Kenneth. *A Genealogy of Modern Architecture: Comparative Critical Analysis of Built Form.* Lars Müller Publishers 2015.

Frowd, Philippe. "The Promises and Pitfalls of Biometric Security Practices in Senegal." *International Political Sociology* 11, no. 4 (2017): 343–359.

Frowd, Philippe. *Security at the Borders. Transnational Practices and Technologies in West Africa.* Cambridge: Cambridge University Press, 2018.

Fuller, Gillian. "Welcome to Windows 2.1: Motion Aesthetics at the Airport." In *Politics at the Airport.* Edited by Mark B. Salter, 161–173. Minneapolis: University of Minnesota Press, 2008.

Galis, Vasilis, Spyros Tzokas, and Aristotle Tympas. "Bodies Folded in Migrant Crypts: Dis/Ability and the Material Culture of Border-Crossing." *Societies* 6, no. 10 (2016): 1–11.

Gammeltoft-Hansen, Thomas, and Nina Nyberg Sørensen. *The Migration Industry and the Commercialization of International Migration.* New York: Routledge, 2013.

Garelli, Glenda, and Martina Tazzioli. "Choucha beyond the Camp: Challenging the Border of Migration Studies." In *The Borders of "Europe": Autonomy of Migration, Tactics of Bordering.* Edited by Nicholas De Genova. 165–184. Durham, NC/London: Duke University Press, 2017.

Gibson, James. *The Ecological Approach to Visual Perception.* Boston: Houghton Mifflin Company, 1979.

Gkliatia, Mariana. "From Frontex to the European Border and Coast Guard: Responsibility for Human Rights Violations at the Borders." Lecture, December 13, 2016, Senate House, London, 2016.

Gkliatia, Mariana, and Herbert Rosenfeldt. "Accountability of the European Border and Coast Guard Agency: Recent Developments, Legal Standards and Existing Mechanisms." School of Advanced Study, University of London. RLI Working Paper No. 30, 2018.

Glouftsios, Georgios. "Designing Digital Borders: The Visa Information System (VIS)." In *Technology and Agency in International Relations*. Edited by Marijn Hoijtink and Matthias Seele. London and New York: Routledge, 2019.

Graham, Stephen. "Flowcity: Networked Mobilities and the Contemporary Metropolis." *Journal of Urban technology* 9, 1 (2002): 4–11.

Gray, John. "Blowing Bubbles." *New York Review of Books*. October 12, 2017.

Green, Jeffrey. *The Eyes of the People: Democracy in an Age of Spectatorship*. Oxford: Oxford University Press, 2011.

Grommé, Francisca. *Governance by Pilot Projects: Experimenting with Surveillance in Dutch Crime Control*. PhD thesis, University of Amsterdam, 2015.

Guild, Elspeth, and Didier Bigo. "The Transformation of European Border Controls." In *Extraterritorial Immigration Control. Legal Challenges*. Edited by Bernard Ryan and Valsamis Mitsilegas. Leiden, Netherlands: Brill Academic Publishers, 2010.

Guldi, Jo. *Roads to Power: Britain Invents the Infrastructure State*. Cambridge, MA: Harvard University Press, 2012.

Gutmann, Amy, and Dennis Thompson. *The Spirit of Compromise: Why Governing Demands It and Campaigning Undermines It*. Princeton, NJ: Princeton University Press, 2010.

Hacking, Ian. *Representing and Intervening: Introductory Topics in the Philosophy of Natural Science*. Cambridge: Cambridge University Press, 1983.

Hajer, Maarten. "The Generic City." *Theory, Culture & Society* 16, no. 4 (1999): 137–144.

Hall, Rachel. *The Transparent Traveler: The Performance and Culture of Airport Security*. Durham, NC/London: Duke University Press, 2015.

Haraway, Donna. "Situated Knowledges: The Science Question in Feminism and the Privilege of Partial Perspective." *Feminist Studies* 14, no. 3 (1988): 575–599.

Harvey, David. "The Political Economy of Public Space." In *The Politics of Public Space*. Edited by Setha Low and Neil Smith. New York: Routledge, 2005.

Hasian, Marouf, Jr. *Forensic Rhetorics and Satellite Surveillance: The Visualization of War Crimes and Human Rights Violations*. New York: Lexington, 2016.

Hecht, Gabrielle. *Entangled Geographies: Empire and Technopolitics in the Global Cold War*. Cambridge, MA: MIT Press, 2011.

Hecht, Gabrielle. "Interscalar Vehicles for an African Anthropocene: On Waste, Temporality, and Violence." *Cultural Anthropology* 33, 1 (2018): 109–141.

Hecht, Gabrielle, and Paul Edwards. "The Technopolitics of Cold War: Toward a Transregional Perspective." In *Essays on Twentieth Century History*. Edited by Michael Adas, 271–314. Philadelphia: Temple University Press, 2010.

Heidegger, Martin. "Building, Dwelling, Thinking." *Basic Writings*. Abingsdon, UK: Routledge, 1993.

Heidegger, Martin. "The Question Concerning Technology." *The Question Concerning Technology and Other Essays*. New York: Harper Perennial, 1977.

Heller, Charles, and Lorenzo Pezzani. "Ebbing and Flowing: The EU's Shifting Practices of (Non-)Assistance and Bordering in a Time of Crisis." *Near Futures Online Issue 1 "Europe at a Crossroads,"* March 2016. http://nearfuturesonline.org/ebbing-and-flowing -the-eus-shifting-practices-of-non-assistance-and-bordering-in-a-time-of-crisis/.

Heller, Charles, Lorenzo Pezzani, and Situ Studio. *Report on the "Left-to-Die-Boat."* Forensic Architecture Project, Goldsmiths University of London, April 11, 2012.

Heller, Charles, Lorenzo Pezzani, and Maurice Stierl. "Disobedient Sensing and Border Struggles at the Maritime Frontier of Europe." *Spheres Media and Migration*, no. 4 (2017): 1–15.

Heller, Charles, and Lorenzo Pezzani. "Liquid Traces. Investigating the Deaths of Migrants at the EU's Maritime Frontier." In *The Borders of "Europe." Autonomy of Migration, Tactics of Bordering*. Edited by Nicholas De Genova, 95–119. Durham, NC/ London: Duke University Press, 2017.

Hernandez, Joel. "Refugee Flows to Lesvos: Evolution of a Humanitarian Response." Migration Policy Institute, January 29, 2016. https://www.migrationpolicy.org /article/refugee-flows-lesvos-evolution-humanitarian-response.

Hernández, Roberto Delgadillo. *Coloniality and Border(ed) Violence: San Diego, San Ysidro and the U-S///Mexico Border*. Tucson: The University of Arizona Press, 2018.

Hoffmann, Josef. *Philosophies of Crime Fiction*. Harpenden/Hertfordshire, UK: No Exit Press, 2013.

Hoijtink, Marijn. *Securing the European "Homeland": Profit, Risk, Authority*. PhD thesis, University of Amsterdam, 2016.

Human Rights Watch. *Subject to Whim: The Treatment of Unaccompanied Children in the French Hautes-Alpes*. New York: Human Rights Watch, 2019.

Idriz, Narin. "The EU-Turkey Deal in Front of the Court of Justice of the EU: An Unsolicited Amicus Brief." T. M. C. Asser Institute for International & European Law, Policy Brief 2017–03, Decenber 1, 2017.

Ingold, Tim. *Being Alive: Essays on Movement, Knowledge and Description.* London: Routledge, 2011.

Institute for Race Relations (IRR). "Humanitarianism: The Unacceptable Face of Solidarity." London: Institute for Race Relations, 2017.

International Organization for Migration (IOM). *Four Decades of Cross-Mediterranean Undocumented Migration to Europe. A Review of the Evidence.* Geneva, Switzerland: International Organization for Migration (IOM), 2017a.

International Organization for Migration (IOM). *World Migration report 2018.* Geneva, Switzerland: International Organization for Migration (IOM), 2017b.

Jacobsen, Katja Lindskov. "On Humanitarian Refugee Biometrics and New Forms of Intervention." *Journal of Intervention and Statebuilding* 11, no. 4 (2017): 529–551.

Janác, Jiří. *European Coasts of Bohemia: Negotiating the Danube-Oder-Elbe Canal in a Troubled Twentieth Century.* Amsterdam: Amsterdam University Press, 2012.

Jansen, Yolande. "Deportability and Racial Europeanization: The Impact of Holocaust Memory and Postcoloniality on the Unfreedom of Movement in and to Europe." In *The Irregularization of Migration in Contemporary Europe: Detention, Deportation, Drowning.* Edited by Yolande Jansen, Robin Celikates, and Joost de Bloois, 15–30. London/New York: Rowman and Littlefield, 2015.

Jansen, Yolande, Robin Celikates, and Joost de Bloois (Eds.). *The Irregularization of Migration in Contemporary Europe.* London/New York: Rowman and Littlefield, 2015.

Jasanoff, Sheila. "Future Imperfect." In *Dreamscapes of Modernity.* Edited by Sheila Jasanoff and Sang Hyun Kim, 1–34. Chicago: University of Chicago Press, 2015.

Jasanoff, Sheila, and Sang-Hyun Kim. "Containing the Atom: Sociotechnical Imaginaries and Nuclear Power in the United States and South Korea." *Minerva* 47, no. 2 (2009): 119–146.

Jeandesboz, Julien. *An Analysis of the Commission Communications on Future Development of Frontex and the Creation of a European Border Surveillance System (Eurosur).* Briefing Paper, European Parliament. Brussels: European Parliament, 2008.

Jeandesboz, Julien. "European Border Policing: EUROSUR, Knowledge, Calculation." *Global Crime* 18, no. 3 (2017): 256–285.

Jones, Reece. *Border Walls: Security and the War on Terror in the United States, India, and Israel.* Chicago: Zed Books, 2012.

Jones, Reece, Corey Johnson, Wendy Brown, Gabriel Popescu, Polly Pallister-Wilkins, Alison Mountz, and Emily Gilbert. "Interventions on the State of Sovereignty at the Border." *Political Geography* 59 (2017): 1–10.

Kapp, Ernst. *Elements of a Philosophy of Technology: On the Evolutionary History of Culture*. Minneapolis: University of Minnesota Press, 2018.

Kasarda, John, and Greg Lindsay. *Aerotropolis. The Way We'll Live Next*. New York: Farrar, Straus and Giroux, 2011.

Kasimis, Charalambos. "Illegal Immigration in the Midst of Crisis." Migration Information Source, March 8, 2012. http://www.migrationpolicy.org/article/greece-illegal -immigration-midst-crisis.

Kasparek, Bernd. "Routes, Corridors, and Spaces of Exception: Governing Migration and Europe." *Near Futures Online* 1 "Europe at a Crossroads" (March 2016). http://nearfuturesonline.org/routes-corridors-and-spaces-of-exception-governing -migration-and-europe/.

Kaufmann, Mareile, and Julien Jeandesboz. "Politics and 'the Digital': From Singularity to Specificity." *European Journal of Social Theory* 20, no. 3 (2016): 309–328.

Kenny, Charles. "Pandemics Close Borders—and Keep Them Closed." *Politico*, March 25, 2020.

Khosravi, Shahram. "Waiting: Keeping Time." In *Migration: A COMPAS Anthology*. Edited by B. Anderson and M. Keith, 72–74. Oxford, UK: COMPAS, 2014.

Knight, Stephen. *Crime Fiction since 1800: Detection, Death, Diversity*. Basingstoke, UK: Palgrave MacMillan, 2010.

Knorr Cetina, Karin, and Alex Preda. "The Temporalization of Financial Markets: From Network to Flow." *Theory, Culture & Society* 24, no. 7 (2007): 116–138.

Knorr Cetina, Karin. "Scopic Media and Global Coordination: The Mediatization of Face-to-Face Encounters." In *Mediatization of Communication*. Edited by Knut Lundby, 39–62. Handbooks of Communication Science Series, Volume 21. Berlin: de Gruyter, 2014.

Koolhaas, Rem, and Bruce Mau. *S, M, L, XL*. Rotterdam, Netherlands/New York: 010 Publishers/Monacelli Press, 1995.

Kotef, Hagar. *Movement and the Ordering of Freedom: On Liberal Governances of Mobility*. Durham, NC/London: Duke University Press, 2015.

Kuchinskaya, Olga. *The Politics of Invisibility: Public Knowledge about Radiation Health Effects after Chernobyl*. Cambridge, MA: MIT Press, 2014.

Kuster, Brigitta, and Vasilis Tsianos. "Erase Them: Eurodac and Digital Deportability." *Transversal Texts*, no. 2 (2013): http://eipcp.net/transversal/0313/kuster-tsianos/en.

Lagendijk, Vincent. *Electrifying Europe: The Power of Europe in the Construction of Electricity Networks*. Amsterdam: Amsterdam University Press, 2008.

Lahav, Gallya, and Vriginie Guiraudon. "Comparative Perspectives on Border Control: Away from the Border and outside the State." In *The Wall around the West: State Borders and Immigration Controls in North America and Europe*. Edited by Peter Andreas and Timothy Snyder, 55–77. Lanham, MD: Rowman and Littlefield Publishers, 2000.

Larkin, Brian. "The Politics and Poetics of Infrastructure." *Annual Review of Anthropology* 42 (2013): 327–343.

Last, Tamara, Giorgia Mirto, Orçun Ulusoy, Ignacio Urquijo, Joke Harte, Nefeli Bami et al. "Deaths at the Borders Database: Evidence of Deceased Migrants' Bodies Found along the Southern External Borders of the European Union." *Journal of Ethnic and Migration Studies* 43, no. 5 (2017): 693–712.

Last, Tamara, and Thomas Spijkerboer. "Tracking Deaths in the Mediterranean." In *Fatal Journeys. Tracking Lives Lost during Migration*. Edited by Tara Brian and Frank Laczko, 85–106. Geneva, Switzerland: International Organization for Migration (IOM), 2014.

Latour, Bruno. *Aramis or the Love of Technology*. Cambridge, MA: Harvard University Press, 1996.

Latour, Bruno. "Drawing Things Together." In *Representation in Scientific Activity*. Edited by Michael Lynch and Steve Woolgar, 19–68. Cambridge, MA: MIT Press, 1990.

Latour, Bruno. *Down to Earth: Politics in the New Climate Regime*. Cambridge: Polity Press, 2018.

Latour, Bruno. *Facing Gaia: Eight Lectures on the New Climate Regime*. Oxford, UK: Wiley-Blackwell, 2017.

Latour, Bruno. "Gabriel Tarde and the End of the Social." In *The Social in Question. New Bearings in History and the Social Sciences*. Edited by Patrick Joyce, 117–132. London: Routledge, 2001.

Latour, Bruno. "Give Me a Laboratory and I Will Raise the World." In *Science Observed. Perspectives on the Social Study of Science*. Edited by Karin Knorr-Cetina and Michael Mulkay, 141–170. London: SAGE, 1983.

Latour, Bruno. *An Inquiry into Modes of Existence*. Cambridge, MA: Harvard University Press, 2013.

Latour, Bruno. "*Onus Orbis Terrarum*: About a Possible Shift in the Definition of Sovereignty." *Millennium: Journal of International Studies* 44, no. 3 (2016): 305–320.

Latour, Bruno. *Pandora's Hope: Essays on the Reality of Science Studies*. Cambridge, MA: Harvard University Press, 1999.

Latour, Bruno. *Paris, Invisible City*, 2006. Accessed May 28, 2020, http://www.bruno -latour.fr/node/343.

Latour, Bruno. *The Pasteurization of France*. Cambridge, MA: Harvard University Press, 1993a.

Latour, Bruno. *Politics of Nature: How to Bring the Sciences into Democracy*. Cambridge, MA: Harvard University Press, 2004.

Latour, Bruno. *Reassembling the Social: An Introduction to Actor-Network-Theory*. Oxford: Oxford University Press, 2005.

Latour, Bruno. *Science in Action*. Cambridge, MA.: Harvard University Press, 1987.

Latour, Bruno. "On Technical Mediation." *Common Knowledge* 3, no. 2 (1994): 29–64.

Latour, Bruno. *We Have Never Been Modern*. Cambridge, MA: Harvard University Press, 1993b.

Latour, Bruno. "When Things Strike Back: A Possible Contribution of Science Studies to the Social Sciences." *British Journal of Sociology* 51, no. 1 (2000): 107–124.

Latour, Bruno. "Why Has Critique Run out of Steam? From Matters of Fact to Matters of Concern." *Critical Inquiry* 30, no. 2 (2004): 225–248.

Latour, Bruno, and Peter Weibel (Eds.). *Making Things Public: Atmospheres of Democracy*. Cambridge, MA: MIT Press, 2005.

Latour, Bruno, and Steve Woolgar. *Laboratory Life: The Social Construction of Scientific Facts*. London: Sage, 1979.

Leander, Anna. "The Politics of Whitelisting: Regulatory Work and Topologies in Commercial Security." *Environment and Planning D: Society and Space* 34, no. 1 (2016): 48–66.

Leerkes, Arjen, and Dennis Broeders. "A Case of Mixed Motives? Formal and Informal Functions of Administrative Immigration Detention." *British Journal of Criminology* 50 (2010): 830–850.

Leong, Sze Tsung. "Gruen Urbanism." In *The Harvard Design School Guide to Shopping*. Edited by Chuihua Judy Chung, Jeffrey Inaba, Rem Koolhaas, and Sze Tsung Leong, 92–128. Cambridge, MA/Cologne, Germany: Harvard Design School and Taschen, 2001.

Leong, Sze Tsung, and Srdjan Jovanovic Weiss. "Air Conditioning." In *The Harvard Design School Guide to Shopping*. Edited by Chuihua Judy Chung, Jeffrey Inaba, Rem Koolhaas and Sze Tsung Leong, 380–390. Cambridge, MA/Cologne, Germany: Harvard Design School and Taschen, 2001.

Leurs, Koen. *Digital Passages: Migrant Youth 2.0. Diaspora, Gender and Youth Cultural Intersections*. Amsterdam: Amsterdam University Press, 2015.

Lommers, Suzanne. *Europe on Air: Interwar Projects for Radio Broadcasting*. Amsterdam: Amsterdam University Press, 2012.

Longo, Matthew. *Politics of Borders. Sovereignty, Security, and the Citizen after 9/11*. Cambridge: Cambridge University Press, 2018.

Lyon, David. *Surveillance after September 11*. Cambridge, UK: Polity Press, 2003.

MacFarlane, Robert. *Mountains of the Mind: Adventures in Reaching the Summit*. New York: Vintage Books, 2004.

Maier, Charles. S. *Once within Borders: Territories of Power, Wealth, and Belonging since 1500*. Cambridge, MA: Harvard University Press, 2016.

Mann, Michael. "The Autonomous Power of the State: Its Origins, Mechanisms and Results." *European Journal of Sociology* 25, no. 2 (1984): 185–213.

Margalit, Avishai. *On Compromise and Rotten Compromises*. Princeton, NJ: Princeton University Press, 2009.

Mazur, Allan. *Energy and Electricity in Industrial Nations: The Sociology and Technology of Energy*. London: Routledge, 2013.

M'charek, Amade, Katharina Schramm, and David Skinner. "Topologies of Race: Doing Territory, Population, and Identity in Europe." *Science, Technology, & Human Values* 39, no. 4 (2014): 468–487.

McLuhan, Marshall. *Understanding Media: The Extensions of Man*. Corte Madera, CA: Gingko Press, 2003.

Meany, Thomas. "A Celebrity Philosopher Explains the Populist Insurgency." *The New Yorker*, February 26, 2018.

Médecins Sans Frontières (MSF). *Harmful Borders: An Analysis of the Daily Struggle of Migrants as they Attempt to Leave Ventimiglia for Northern Europe*. Milan/Rome: Médecins Sans Frontières, 2018.

Meiches, Benjamin. "Non-human Humanitarians." *Review of International Studies* 45, no. 1 (2019): 1–19.

Mezzadra, Sandro, and Brett Neilson. *Border as Method, or the Multiplication of Labor*. Durham, NC/London: Duke University Press, 2013.

Mirzoeff, Nicholas. *The Right to Look: A Counterhistory of Visuality*. Durham, NC/London: Duke University Press, 2011.

Misa, Thomas, and Johan Schot. "Introduction." *History and Technology* 21, no. 1 (2005): 1–19.

Mitchell, A. Wess, and Charles Ingrao, "Emperor Joseph's Solution to Coronavirus." *Wall Street Journal*, April 6, 2020.

Mitchell, Timothy. *Carbon Democracy: Political Power in the Age of Oil*. London/New York: Verso, 2011.

Mitchell, Timothy. *Rule of Experts: Egypt, Technopolitics, Modernity*. Berkeley: University of California Press, 2002.

Mol, Annemarie. "Actor-Network Theory: Sensitive Terms and Enduring Tensions." *Kölner Zeitschrift für Soziologie und Sozialpsychologie* 50, no. 1 (2010): 253–269.

Monar, Jörg. "The Dynamics of Justice and Home Affairs: Laboratories, Driving Factors and Costs." *Journal of Common Market Studies* 39, no. 4 (2001): 747–764.

Mountz, Alison. "The Enforcement Archipelago: Detention, Haunting, and Asylum on Islands." *Political Geography* 30 (2011): 118–128.

Mountz, Alison. "Political Geography II: Islands and Archipelagos." *Progress in Human Geography* 39, no. 5 (2014): 636–646.

Mountz, Alison. *Seeking Asylum: Human Smuggling and Bureaucracy at the Border*. Minneapolis: University of Minnesota Press, 2010.

Nail, Thomas. *Theory of the Border*. New York: Oxford University Press, 2016.

Netz, Reviel. *Barbed Wire: An Ecology of Modernity*. Middletown, CT: Wesleyan University Press, 2004.

Nikolaeva, Anna. "Designing Public Space for Mobility: Contestation, Negotiation and Experiment at Amsterdam Airport Schiphol." *Tijdschrift voor Economische en Sociale Geografie* 103, no. 5 (2012): 542–554.

Nyers, Peter. "Moving Borders: The Politics of Dirt." *Radical Philosophy* 174 (July/August 2012): 2–6.

Öktem, Kerem. "The Ambivalent Sea: Regionalizing the Mediterranean Differently." In *Mediterranean Frontiers. Borders, Conflict and Memory in a Transnational World*. Edited by Dimitar Bechev and Kalypso Nicolaidis. London: New York: Tauris Academic Studies, 2010.

Oosterling, Henk. "Dasein als design." *De Groene Amsterdammer* 130, no. 14 (2009): 32–36.

Opitz, Sven, and Uwe Telemann. "Europe as Infrastructure: Networking the Operative Community." *South Atlantic Quarterly* 141, no. 1 (2015): 171–190.

Pallister-Wilkins, Polly. "How Walls Do Work: Security Barriers as Devices of Interruption and Data Capture." *Security Dialogue* 47, no. 2 (2016): 151–164.

Pallister-Wilkins, Polly. "Humanitarian Borderwork: Actors, Spaces, Categories." In Reece Jones, Corey Johnson, Wendy Brown, Gabriel Popescu, Polly Pallister-Wilkins, Alison Mountz and Emily Gilbert. "Interventions on the State of Sovereignty at the Border." *Political Geography* 59 (2017): 1–10.

Pallister-Wilkins, Polly. "The Humanitarian Politics of European Border Policing: Frontex and Border Police in Evros." *International Political Sociology* 9, no. 1 (2015): 53–69.

Papataxiarchis, Evthymios. "Being 'There': At the Front Line of the 'European Refugee Crisis'—Part 1." *Anthropology Today* 32, no. 2 (2016): 5–9.

Papataxiarchis, Evthymios. "Being 'There': At the Front Line of the 'European Refugee Crisis'—Part 2." *Anthropology Today* 32, no. 3 (2016): 3–7.

Passoth, Jan Hendrik, and Nicholas Rowland. "Actor-Network State: Integrating Actor-Network Theory and State Theory." *International Sociology* 25, no. 6 (2010): 818–841.

Pelizza, Annalisa. "Processing Alterity, Enacting Europe: Migrant Registration and Identification as Co-construction of Individuals and Polities." *Science, Technology, & Human Values* 25, no. 2 (2019): 262–288.

Pollozek, Silvan and Jan Hendrik Passoth. "Infrastructuring European Migration and Border Control: The Logistics of Registration and Identification at Moria Hotspot." *EPD: Society and Space* 37, no. 4 (2019): 606–624.

Polman, Linda. "Tegen elke prijs." *De Groene Amsterdammer* 40 (2020): 42–44.

Ponzanesi, Sandra. "Digital Diasporas: Postcoloniality, Media and Affect." *Interventions*, February 6, 2020. https:doi.org/10.1080/1369801X.2020.1718537.

Priestman, Martin (Ed.). *The Cambridge Companion to Crime Fiction*. Cambridge: Cambridge University Press, 2010.

Pritchard, Sara B. *Confluence: The Nature of Technology and the Remaking of the Rhône*. Cambridge, MA: Harvard University Press, 2011.

Proctor, Robert. "Agnotology: A Missing Term to Describe the Cultural Production of Ignorance (and Its Study)." In *Agnotology: The Making and Unmaking of Ignorance*. Edited by Robert Proctor and Londa Schiebinger, 1–33. Stanford, CA: Stanford University Press, 2008.

Rancière, Jacques. *Dissensus: On Politics and Aesthetics*. London: Bloomsbury, 2010.

Rancière, Jacques. *The Politics of Aesthetics: The Distribution of the Sensible*. London/New York: Continuum, 2004.

Roberts, Stephen, and Stefan Elbe. "Catching the Flu: Syndromic Surveillance, Algorithmic Governmentality and Global Health Security." *Security Dialogue* 48, no. 1 (2017): 46–62.

Rodgers, Dennis, and Bruce O'Neill. "Infrastructural Violence: Introduction to the Special Issue." *Ethnography* 13, no. 4 (2012): 401–412.

Rosanvallon, Pierre. *Counter-Democracy: Politics in an Age of Distrust*. Cambridge: Cambridge University Press, 2008.

Rowland, Nicholas, and Jan-Hendrik Passoth. "Infrastructure and the State in Science and Technology Studies." *Social Studies of Science* 45, no. 1 (2015): 137–145.

Rozakou, Katerina. "Socialities of Solidarity: Revisiting the Gift Taboo in Times of Crises." *Social Anthropology* 24, no. 2 (2016): 185–199.

Ruhrmann, Henriette, and David FitzGerald. *The Externalization of Europe's Borders in the Refugee Crisis, 2015–2016*. San Diego: University of California Center for Comparative Immigration Studies, 2017.

Rumford, Chris. *Citizens and Borderwork in Contemporary Europe*. London: Routledge, 2008.

Ryan, Barry. "Security Spheres: A Phenomenology of Maritime Spatial Practices." *Security Dialogue* 46, no. 6 (2015): 568–584.

Said, Edward. *Orientalism: Western Conceptions of the Orient*. London: Penguin, 1977.

Salter, Mark B. "The Global Visa Regime and the Political Technologies of the International Self: Borders, Bodies, Biopolitics." *Alternatives: Global, Local, Political* 31, no. 2 (2006): 167–189.

Salter, Mark B. (Ed.). *Making Things International 1: Circuits and Motion*. Minneapolis/London: University of Minnesota Press, 2015.

Salter, Mark B. (Ed.). *Making Things International 2: Catalysts and Reactions*. Minneapolis/London: University of Minnesota Press, 2016.

Salter, Mark B. (Ed.). *Politics at the Airport*. Minneapolis: University of Minnesota Press, 2008.

Salter, Mark B., and William Walters. "Bruno Latour Encounters International Relations: An Interview." *Millennium: Journal of International Studies* 44, no. 3 (2016): 534–546.

Scaggs, John. *Crime Fiction*. London/New York: Routledge, 2005.

Scheel, Stephan. *Autonomy of Migration? Appropriating Mobility Within Biometric Border Regimes*. London/New York: CRC Press, 2019.

Schinkel, Willem, and Liesbeth Noordegraaf-Eelens. *In Medias Res: Peter Sloterdijk's Spherological Poetics of Being*. Amsterdam: Amsterdam University Press, 2011.

Schipper, Frank, and Johan Schot. "Infrastructural Europeanism, or the Project of Building Europe on Infrastructures: An Introduction." *History and Technology* 27, no. 3 (2011): 245–264.

Schouten, Peer. "Security as Controversy: Reassembling Security at Amsterdam Airport." *Security Dialogue* 45, no. 1. (2014): 23–42.

Schueler, Judith. *Materialising Identity: The Co-construction of the Gotthard Railway and Swiss National Identity*. Eindhoven, Netherlands: Technische Universiteit Eindhoven, 2008.

Scott, James C. *Seeing Like a State: How Certain Schemes to Improve the Human Condition Have Failed.* New Haven, CT/London: Yale University Press, 1998.

Self, Will. "The Frowniest Spot on Earth." *London Review of Books* 33, no. 9 (2011): 10–11.

Serres, Michel. *Eyes.* London: Bloomsbury, 2015.

Sloterdijk, Peter. *Regeln für den Menschenpark. Ein Antwortschreiben zu Heideggers Brief über den Humanismus.* Frankfurt: Suhrkamp, 2008.

Sloterdijk, Peter. *Bubbles.* Cambridge, MA: MIT Press, 2011.

Sloterdijk, Peter. *Falls Europa erwacht: Gedanken zum Programm einer Weltmacht am Ende des Zeitalters ihrer politischen Absence.* Berlin: Suhrkamp, 1994.

Sloterdijk, Peter. *Foams.* Cambridge, MA: MIT Press, 2016.

Sloterdijk, Peter. *Globes.* Cambridge, MA: MIT Press, 2014.

Sloterdijk, Peter. *Infinite Mobilization.* Cambridge, UK: Polity Press, 2020.

Sloterdijk, Peter. *Not Saved: Essay after Heidegger.* Cambridge, UK: Polity Press, 2017.

Sloterdijk, Peter. *Selected Exaggerations: Conversations and Interviews 1993–2012.* Edited by B. Klein. Cambridge, UK: Polity, 2016.

Sloterdijk, Peter, and H.-J. Heinrichs. *Die Sonne und der Tod: Dialogische Untersuchungen.* Frankfurt am Main, Germany: Suhrkamp, 2001.

Sloterdijk, Peter, and Gesa Mueller von der Hagen. "Instant Democracy: the Pneumatic Parliament." In *Making Things Public: Atmospheres of Democracy.* Edited by Bruno Latour and Peter Weibel, 952–957. Cambridge, MA: MIT Press, 2005.

Smets, Kevin. "The Way Syrian Refugees in Turkey Use Media: Understanding 'Connected Refugees' through a Non-mediacentric and Local Approach." *Communications* 43, no. 1 (2018):113–123.

Sontowski, Simon. "Speed, Timing and Duration: Contested Temporalities, Techno-political Controversies and the Emergence of the EU's Smart Border." *Journal of Ethnic and Migration Studies* 44, no. 16 (2018): 2730–2746.

Spijkerboer, Thomas. "Afterword: From the Iron Curtain to Lampedusa." In *Border Deaths: Causes, Dynamics and Consequences of Migration-related Mortality.* Edited by Paolo Cuttitta and Tamara Last. Amsterdam: Amsterdam University Press, 2020.

Spijkerboer, Thomas. "The Global Mobility Infrastructure: Reconceptualising the Externalisation of Migration Control." *European Journal of Migration and Law* 20, no. 4 (2018): 452–469.

Squire, Vicky. "Desert 'Trash': Posthumanism, Border Struggles, and Humanitarian Politics." *Political Geography* 39 (2014): 11–21.

Stäheli, Urs. "Indexing—The Politics of Invisibility." *Environment and Planning D: Society and Space* 34, no. 1 (2016): 14–29.

Stengers, Isabelle. *The Invention of Modern Science.* Minneapolis/London: University of Minnesota Press, 2000.

Strathern, Marilyn. *The Relation: Issues in Complexity and Scale.* Cambridge, UK: Prickly Pear Press, 1995.

Strik, Tineke. "Parliamentary Assembly of the Council of Europe." *Lives Lost in the Mediterranean Sea: Who is Responsible?* Doc. 12895, April 5, 2012.

Suchman, Lucy. "Situational Awareness: Deadly Bioconvergence at the Boundaries of Bodies and Machines." *Media Tropes* 5, no. 1 (2015): 1–24.

Tamma, Paola, and Hannie Cokelaere. "Schengen Proves Hard to Reboot after System Meltdown." *Politico.* Accessed June 1, 2020, https://www.politico.eu/article/schengen-proves-hard-to-reboot-after-system-meltdown/.

Taylor, Linnet. "No Place to Hide? The Ethics and Analytics of Tracking Mobility Using Mobile Phone Data." *Environment and Planning D: Society and Space* 34, no. 2 (2016): 319–336.

Tazzioli, Martina. "Containment through Mobility: Migrants' Spatial Disobediences and the Reshaping of Control through the Hotspot System." *Journal of Ethnic and Migration Studies* 44, no. 16 (2018): 2764–2779.

Tazzioli, Martina. "Eurosur, Humanitarian Visibility, and (Nearly) Real-Time Mapping in the Mediterranean." *ACME: An International Journal for Critical Geographies* 15 no. 3 (2016): 561–579.

Tazzioli, Martina. "Spy, Track and Archive: The Temporality of Visibility in Eurosur and Jora." *Security Dialogue* 49, no. 4 (2018): 272–288.

Ten Bos, René. "Towards an Amphibious Anthropology: Water and Peter Sloterdijk." *Environment and Planning D: Society and Space* 27 (2009): 73–86.

Topak, Özgün. "The Biopolitical Border in Practice: Surveillance and Death at the Greece–Turkey Borderzones." *Environment and Planning D: Society and Space* 32, no. 5 (2014): 815–833.

Torpey, John. *The Invention of the Passport. Surveillance, Citizenship and the State.* Cambridge: Cambridge University Press, 1998.

Uçarer, Emek. "Justice and Home Affairs." In *European Union Politics.* Edited by Michelle Cini and Nieves Pérez-Solórzano Borragán. Oxford: Oxford University Press, 2019.

Urbinati, Nadia. *Democracy Disfigured: Opinion, Truth, and the People.* Cambridge, MA: Harvard University Press, 2014.

Van den Eede, Yoni. "In between Us: On the Transparency and Opacity of Techno-logical Mediation." *Foundations of Science* 16, no. 2–3 (2011): 139–159.

Van Houtum, Henk. "Human Blacklisting: The Global Apartheid of the EU's Exter-nal Border Regime." *Environment and Planning D: Society and Space* 28, no. 6 (2010): 957–976.

Van der Vleuten, Erik, and Arne Kaijser. "Networking Europe." *History and Technol-ogy* 21 no. 1 (2005): 21–48.

Van Reekum, Rogier, and Willem Schinkel. "Drawing Lines, Enacting Migration: Visual Prostheses of Bordering Europe." *Public Culture* 29, no. 1 (81) (2017): 27–51.

Van Tuinen, Sjoerd. *Sloterdijk. Binnenstebuiten denken*. Kampen, Netherlands: Klement/ Pelckmans, 2004.

Vaughan-Williams, Nick. *Border Politics: The Limits of Sovereign Power*. Edinburgh: Edinburgh University Press, 2009.

Vaughan-Williams, Nick. "Borderwork beyond Inside/Outside? Frontex, the Citizen-Detective and the War on Terror." *Space and Polity* 12, no. 1 (2008): 63–79.

Verbeek, Peter-Paul. *What Things Do: Philosophical Reflections on Technology, Agency, and Design*. University Park: Pennsylvania State University Press, 2005.

Voegelin, Eric. *The New Science of Politics*. Chicago: University of Chicago Press, 1987.

Vradis, Antonis, Evie Papada, Joe Painter, and Anna Papoutsi. *New Borders: Hotspots and the European Migration Regime*. London: Pluto Press, 2019.

Walker, Rob. *Inside/Outside: International Relations as Political Theory*. Cambridge: Cambridge University Press, 1993.

Walters, William. "Drone Strikes, Dingpolitik and Beyond: Furthering the Debate on Materiality and Security." *Security Dialogue* 45, no. 2 (2014): 101–118.

Walters, William. "Foucault and Frontiers: Notes on the Birth of the Humanitarian Border." In *Governmentality: Current Issues and Future Challenges*. Edited by Ulrich Bröckling, Suzanne Krasmann and Thomas Lemke, 138–164. New York: Routledge, 2011.

Walters, William. "Live Governance, Borders, and the Time–Space of the Situation: EUROSUR and the Genealogy of Bordering in Europe." *Comparative European Politics* 15, no. 5 (2017): 794–817.

Walters, William. "Migration, Vehicles, and Politics: Three Theses on Viapolitics." *European Journal of Social Theory* 18, no. 4 (2015): 469–488.

Walters, William. "Rezoning the Global: Technological Zones, Technological Work, and the (Un-)making of Biometric Borders." In *The Contested Politics of Mobility: Bor-derzones and Irregularity*. Edited by Vicky Squire, 51–73. London: Routledge, 2011.

Walters, William, and Anne-Marie D'Aoust. "Bringing Publics into Critical Security Studies: Notes for a Research Strategy." *Millennium: Journal of International Studies* 44, no. 1 (2015): 45–68.

Walters, William, and Jens Henrik Haahr. *Governing Europe: Discourse Governmentality and European Integration.* New York: Routledge, 2005.

Weizman, Eyal. *Forensic Architecture. Violence at the Threshold of Detectability.* New York: Zone Books, 2017.

Xiang, Biao, and Johan Lindquist. "Migration Infrastructure." *International Migration Review* 48, no. 1 (2014): 122–148.

Zaiotti, Ruben. *Cultures of Border Control. Schengen and the Evolution of European Frontiers.* Chicago: University of Chicago Press, 2011.

Zolberg, Aristide R. "The Archaeology of 'Remote Control.'" In *Migration Control in the North Atlantic World: The Evolution of State Practices in Europe and the United States from the French Revolution to the Inter-War Period.* Edited by Andreas Fahrmeier, Oliver Faron, and Patrick Weil, 195–221. New York: Berghahn Books, 2003.

Index